Design-Build Explained

Macmillan Building and Surveying Series
Series Editor: IVOR H. SEELEY
Emeritus Professor, The Nottingham Trent University

Advanced Building Measurement, second edition Ivor H. Seeley
Advanced Valuation Diane Butler and David Richmond
An Introduction to Building Services Christopher A. Howard
Applied Valuation Diane Butler
Asset Valuation Michael Rayner
Building Economics, third edition Ivor H. Seeley
Building Maintenance, second edition Ivor H. Seeley
Building Procurement Alan E. Turner
Building Quantities Explained, fourth edition Ivor H. Seeley
Building Surveys, Reports and Dilapidations Ivor H. Seeley
Building Technology, fourth edition Ivor H. Seeley
Civil Engineering Contract Administration and Control
    Ivor H. Seeley
Civil Engineering Quantities, fifth edition Ivor H. Seeley
Civil Engineering Specification, second edition Ivor H. Seeley
Computers and Quantity Surveyors Adrian Smith
Contract Planning and Contractual Procedures, second edition
    B. Cooke
Contract Planning Case Studies B. Cooke
Design-Build Explained David E. L. Janssens
Development Site Evaluation N. P. Taylor
Environmental Science in Building, second edition R. McMullan
Housing Associations Helen Cope
Housing Management: Changing Practice edited by Christine Davies
Information Technology Applications in Commercial Property edited by
    Rosemary Feenan and Tim Dixon
Introduction to Valuation D. Richmond
Marketing and Property People Owen Bevan
Principles of Property Investment and Pricing W. D. Fraser
Property Valuation Techniques David Isaac and Terry Steley
Quality Assurance in Building Alan Griffith
Quantity Surveying Practice Ivor H. Seeley
Structural Detailing, second edition P. Newton
Urban Land Economics and Public Policy, fourth edition
    P. N. Balchin, J. L. Kieve and G. H. Bull
Urban Renewal – Theory and Practice Chris Couch
1980 JCT Standard Form of Building Contract, second edition
    R. F. Fellows

Series Standing Order
If you would like to receive future titles in this series as they are published, you can
make use of our standing order facility. To place a standing order please contact your
bookseller or, in case of difficulty,write to us at the address below with your name
and address and the name of the series. Please state with which title you wish to
begin your standing order. (If you live outside the United Kingdom we may not have
the rights for your area, in which case we will forward your order to the publisher
concerned.)

Customer Services Department, Macmillan Distribution Ltd
Houndmills, Basingstoke, Hampshire RG21 2XS, England

Design–Build
Explained

David E.L. Janssens
BSc (Eng), CEng, MICE

MACMILLAN

First published 1991 by
THE MACMILLAN PRESS LTD
Houndmills, Basingstoke, Hampshire RG21 2XS
and London
Companies and representatives
throughout the world

Cover photograph credits:
Client: St Martin's Property Corporation
Design–Build Contractor: Tarmac Construction Ltd
Architect: Fitch Benoy
Photograph: David Janssens

ISBN 0–333–54810–8 hardcover
ISBN 0–333–54811–6 paperback

A catalogue record for this book is available
from the British Library.

Printed in Hong Kong

Reprinted 1993

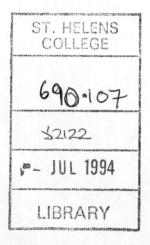

Contents

Check Lists, Tables and Charts

Preface

For some thirty years I have been involved in Design–Build contracting, in one form or another, starting with my first job as an indentured civil engineer on an oil storage installation in Purfleet in Essex. The site was run by engineers and as is common on civil engineering projects no one wanted to look after the buildings. After just a few months on site, the site agent, a fearsome individual to me at least, called me into his office and told me that, from then on, I would be in charge of the buildings. I could not have managed but for the assistance of the foreman carpenter whose name, I remember, was Cecil. As a site engineer, come assistant agent, I went on from there to two or three other building sites, and it was during this period that I developed my interest in building work, as opposed to civil engineering. It was a Design–Build contract and I enjoyed the open relationship with the design team: we were working to the same goals.

At another stage in my career, I hasten to add *before* I joined my present employer, I was the project manager on a large project under the traditional UK form of contract. The programme was tight, and we had made a number of mistakes in the tender. As I remember it, we had priced for some five 22RB mobile cranes for about six months, but had forgotten to price in the drivers, and we had done the same with four hoists. We had omitted to allow for chainmen to work for the six site engineers. All-in-all, these and other omissions not only wiped out our potential profit, but also they could have involved us in a substantial loss.

The remedy, not open to a Design–Build contractor, was to throw up an immediate smoke screen, and 'use' the Contract. When I say that we finished the contract some nine months late, without incurring damages for delay, but on the contrary succeeding with a claim which not only restored the tendered profit but enhanced it considerably, you will realise how effectively, in a Traditional Contract, a contractor can make use of shortcomings of the employer's consultants to overcome the effect of his own mistakes.

Whilst I obtained a degree of satisfaction from this experience and achievement, it was not my style. For me the team approach was the way to get real job satisfaction from contracting, and I am not alone in this.

I come now to the reasons why I decided to give up virtually every week-end, and most of my holiday periods, for a year, to devote the time to write this book.

There were two distinct and separate reasons; one arose from a conversation with an unknown student at Wolverhampton Polytechnic, and the other came out of sheer frustration and annoyance with both the technical and lay press with its coverage of design-build issues: their ignorance has troubled me.

To enlarge upon this, I cite the instance of phase 1 of the BBC headquarters at the White City in London. The architectural, and largely the building industry's press, deplored the idea that the proposed headquarters should be built under a Design–Build form of contract. It was reported, and is probably true, that the Royal Fine Arts Commission condemned the idea.

But I ask, what business is it of the press or the Royal Fine Arts Commission to condemn a form of contracting *which has no bearing whatever on the final design and appearance of the building,* but which goes a long way to ensure that the risk of cost escalation, delay and claims for extension of time, and loss and expense, by the contractor are minimised, so saving the license fee, and tax-payers' money?

What these bodies fail to appreciate is, that the BBC headquarters was designed in shape, form, layout and appearance by architects employed by the BBC themselves. They could have commissioned any designer in the world to do it for them.

In fact, if the BBC had so wished, they could have issued documents in the form of room schedules, spatial relationships and performance criteria to selected Design–Build contractors teamed-up with distinguished architects inviting proposals from them. This would have been, in effect, a *costed* design competition.

So, it was not the form of contract which determined the design of that building, but the employer's architect, and so the debate has nothing whatever to do with Design–Build. It is about the merits of the building design itself. This book has been written for those, among others, who do not understand this point.

The other occurrence that spurred me to write this book came, as I said, from an unknown student at Wolverhampton Polytechnic. After I gave a talk at the Polytechnic on Design–Build, concentrating on how it actually works in practice, this student told me that he had access to commentaries on the standard forms of contract, practice notes and various other academic works and contractors' brochures, but none quite gave the information in the way that my talk was structured. He asked me for a copy of my notes. That was impractical, as they were too brief, not to say unintelligible, but I did recognise from his request the need among the students of our industry for as much practical help and guidance as possible to enable them to understand the context in which they will be working when they leave college.

The standard of the practical side of technical teaching has come a long way since the days when I was at university. My course was so academic, that when I started work after graduating, I was ignorant of such basics as the standard size of a brick, how to detail and schedule reinforcement, what a setting out profile was, and many other such examples. We studied some basics of law as an extra subject, but had no tuition on building or civil engineering forms of contract.

We, who work in the industry can help our students still further, and I hope by this work I have made some contribution to this cause.

As a footnote, I add that any views or opinions expressed in this book are my own, and not those of my employer.

David E. L. Janssens

Acknowledgements

There are so many associates, colleagues and ex-colleagues who have helped me over the years in the development of my knowledge and understanding of Design–Build contracting, that it would be invidious, if not impossible to single out a few to mention, so I thank them collectively. Those who have been closest to me know who they are, and I thank them most sincerely.

There are a number of other special people who have helped me, either directly or indirectly, in writing this book, and I shall mention them by name.

My son Mark, who is a chartered quantity surveyor in private practice, graduated, with honours from Trent Polytechnic, and it was through him, when he was a student there, that my first talk to students on Design–Build was arranged. He has also helped me in a more direct way with the book, both in his comments and contributions and as a sounding board for ideas.

My son Stephen, with his honours degree in Latin, has helped me with the considerable problems that I have experienced with my rusty English grammar.

Dominic, my other son, also an honours graduate (in civil engineering), now turned banker, must have a mention. He has been away until recently and so could not help directly, but I know that he has been with me in spirit and that is enough for me.

Tamara and Teresa, my two daughters, are a great source of joy and fun. They have certainly prevented me from becoming pompous. Without the humour about the house it would have been impossible to remain sane, spending so much time at the word-processor. And, incidentally, I thank them for not yawning or showing boredom, no matter how they must have felt, every time I mentioned *The Book*.

Then, and most important of all, there is my dear wife, Marge. I cannot say enough to thank her, not just for the tea and coffee on demand, and at the right time a little aperitif, but also for her forbearance with unorthodox meal-times, for taking over much of the administration of household affairs which I have temporarily relinquished and for being, this year, the master gardener and labourer in the continued creation of the garden of our recently, self-design-built house.

I Basic Concepts

1 Introduction and Reading Routes

'What! Have you not read it through?
No Sir; Do you read books through'

Johnson   [Boswell's Life of Johnson]

KEY TOPICS

- Who the book is aimed at.
- Reading routes for different interests.
- Other publications to refer to.

INTENDED READERSHIP

This book is intended for:

Students

To supplement their academic studies, by giving them an insight into the practical workings of Design–Build, as a 'living', flexible, changing method of procuring construction work.

Employers

To give them confidence in choosing the Design–Build route, perhaps for the first time, and helping them to select the Design–Build variety best suited to their project. Also for those who have previously used Design–Build with limited success, to help them identify the causes, and overcome them on a future project.

Architects and Engineers

To help them to understand what both employers and contractors want from their consultants in the Design–Build situation.

Quantity Surveyors

To assist them in advising employers and to help them identify the role that the quantity surveyors can play in Design–Build.

Contractors

To help them identify the considerable additional risks that they incur in Design–Build, and how to manage those risks.

READING ROUTES

Recognising that most people have insufficient time to read a book, such as this, continuously from cover to cover, the following reading routes will assist selective readers, or re-readers, to concentrate on the topics of most interest to them.

Students

Students should read, or at least scan, the whole book, then re-read any section which is of particular interest to them as a topic in their syllabus, or perhaps as a topic to form the basis of a dissertation or 'project'.

Employers, with an Impending Project

It is suggested that employers read through the book in the following sequence, with the noted degree of importance:

Chapter		Comment
2	Glossary of terms	Essential
3	Recent development of Design–Build	Useful
4	Forms of contract	Useful
5	Varieties of Design–Build procurement	Essential
6	Which form of procurement	Essential
7	Employer's representation	Essential
8	Employer's consultants, appointment terms	Essential
9	Contractor selection	Useful
19	Collateral warranties	Useful
10	Tender evaluation	In due course
11–16	Contractor's view	Of interest
17	Valuation of changes	Essential
18	Value engineering	Useful

Employers, with a General Interest

Employers who do not have an imminent project, but wish to obtain a basic understanding of Design–Build, are recommended to scan through the whole book, concentracting on the aspects which interest them most. However, it is recommended that Chapter 5, in particular, is read reasonably carefully.

Architects, Engineers and Quantity Surveyors

It is recommended that Consultants read through the following chapters, with the noted degree of importance:

Chapter		*Comment*
2	Glossary of terms	Essential
3	Recent development of Design–Build	Useful
4	Forms of contract	Useful
5	Varieties of Design–Build procurement	Essential
6	Which form of procurement	Useful
	Employer's consultants	
7	Employer's representation	Essential
8	Employer's consultants, appointment terms	Essential
9	Contractor selection	Essential
10	Tender evaluation	Useful
	Contractor's consultants	
11	Contractor's responsibilities/risks	Useful
12	Contractor's consultants	Essential
13	...pre-contract services	Essential
14	Tendering	Essential
15	Contract documentation	Essential
16	...post-contract services	Essential
17	Valuation of changes	Useful
19	Collateral warranties	Essential

Contractors

The reading routes for contractors' personnel will vary depending upon the reader's position, and hence the guidance given is very general:

Chapter		*Comment*
3	Recent development of Design–Build	Useful
4	Forms of contract	Essential
5	Varieties of Design–Build procurement	Essential
11–17	Contractor's view	Essential
6–10	Employer's view	Useful
18	Value engineering	Useful
19	Collateral warranties	Essential

REFERENCE PUBLICATIONS

To understand the contents of this book fully it is essential to have access to some or all of the following publications, depending upon one's interest.

JCT Publications

Obtainable from most institution bookshops, and by post from

> RIBA Publications
> Finsbury Mission, 39 Moreland St
> London EC1V 8BB   Tel 071 251 0791

> Standard Forms of Building Contract With Contractor's Design, 1981
> Practice Note CD/1A
> Practice Note CD/1B
> The Use of Standard Forms of Building Contract

BPF Publications

Obtainable by post from

> The British Property Federation Ltd.
> 35 Catherine Place
> London SW1E 6DY   Tel 071 828 0111

> Manual of the BPF System
> BPF Edition ACA Form of Building Agreement, 1984

HMSO Publications

Obtainable from HMSO bookshops, HMSO accreditted agents, and by post from

> HMSO Publications Centre
> PO Box 276, London SW8 5DT   Tel 071 873 9090

> General Conditions of Contract for Building & Civil Engineering (GC/WORKS/1 – Edition 3)
> DHSS CAPRICODE Health Building Procedure

RIBA Publications

Obtainable from the RIBA, 66 Portland Place, London, or by post from

> RIBA Publications, address as above.
> Architect's Appointment

The Association of Consulting Engineers Publications

Obtainable from institution bookshops, or by post from

> The Association of Consulting Engineers
> Alliance House 12 Caxton Street
> Westminster SW1H 0QL   Tel 071 222 6557

> ACE Conditions of Engagement
> Agreements 1, 2, 3, 4A (i), (ii), (iii) and 4B

The Royal Institution of Chartered Surveyors Publications

Obtainable from

> The Royal Institution of Chartered Surveyors
> 12 Great George Street
> London SW1P 3AD

> Standard Form of Agreement for the Appointment of a Quantity Surveyor
> Standard Conditions of Engagement
> Guidance Notes for Completing the Appendix

2 Definitions

There are many terms used in the construction industry which do not have universally agreed meanings, and this is particularly true in relation to Design–Build. Most such terms are described fully in the body of the book, but in an endeavour to avoid confusion at the outset this glossary has been included to describe the meaning which the author has ascribed to the terms.

Throughout the Book, capitals have been used to denote specific systems, documents, etc. For example, Traditional Contracting is the system described in this glossary and is not any traditional system; Contract Sum Analysis is a specific JCT 1981 contract document and not merely an analysis of a contract sum. Occasionally a capital is used for Employer or Contractor to stress that they are specific parties to the Contract under discussion.

Architect:
> For the purpose of this book, an architect is one who undertakes architectural design, although it is recognised that such a practitioner is not entitled to use this title unless he is registered.

Bill of Quantities:
> A detailed list of the quantities and description of the work involved in the construction of the Works. It does not necessarily have to conform to a pre-determined format, such as the Standard Method of Measurement.

Building:
> That which the Employer contracts with the Contractor to build; synonymous with Project and Works.

Client:
> The firm who enters into a contract with a contractor for the construction of the Works. The word is synonymous with Employer, as in the JCT form and Authority as in the GC/WORKS/1 form.

Consultant:
> Any independent designer, advisor or quantity surveyor employed by a contractor or employer.

Contractor:
> The firm which undertakes the (design and) construction of the Works.

Contractor's Proposals:
  With capital C and P this is a JCT 1981 contract document.
Cost Plan:
  Similar to a bill of quantities, but less detailed, and consisting mainly of elements of work against which lump-sum or composite prices are put to arrive at an estimated cost for the Works.
Design:
  A general term which includes the drawings and specifications which define the scope and nature of the works which are to be constructed.
Design–build:
  A form of building procurement whereby the Contractor who constructs the Works, also undertakes all of, or a proportion of, the design of the Works.
Design and Manage:
  A form of building procurement whereby the Contractor is responsible for the design and the management of the construction of the Works undertaken by a series of works contractors. There are two forms of Design and Manage; Design–Build or Management Contracting, depending on the way in which the Contract Sum is established. See Chapter 5.
Develop and Construct:
  A variety of Design–Build whereby the Employer's consultants complete a substantial proportion of the design of the Works. See Chapter 5.
Employer's Requirements:
  With a capital E and R, this is a JCT 1981 contract document.
JCT 81 Form:
  Abbreviation for the JCT Standard Form of Building Contract with Contractor's Design.
Negotiated D–B Contract:
  In the context of this book the term is used to describe only those negotiated contracts in which the Contractor undertakes the design of the Works from the outset.
Single-Stage Tendering, Design–Build:
  A form of building procurement whereby the Employer's consultants undertake the conceptual design and establish performance criteria, and the Contractor completes the detailed design; similar to Develop and Construct, but the Contractor is given more scope within which to work. See Chapter 5.
Traditional Contracting:
  A method of building procurement, traditional in the UK, where responsibility for the design lies with the Employer's consultants. The work is valued in accordance with a bill of quantities prepared by the Employer's quantity surveyor and priced by the Contractor and the Contractor is responsible for construction and not design. See Chapter 6.

Turnkey Contract:

A form of building procurement whereby the Contractor undertakes the design, construction and equipping of the project, usually to a defined overall performance specification. It is usually applied to major industrial projects. See Chapter 5.

Two-Stage Tendering, Design–Build:

A form of building procurement whereby the Contractor is chosen on the basis of an offer which is incomplete in some way, and then in a second stage the Contractor negotiates the Contract with the Employer. See Chapter 5.

3 Recent Development of Design–Build

How very little, since things were made,
things have altered in the building trade.

Kipling   [A Truthful Song]

KEY TOPICS

- The origins of Design–Build
- Package Deals
- Stagnatinn of Package Deal business
- Design–Build of the 1980s and 1990s
- Changing character of the construction industry
- Publication of the JCT 1981 Form
- Resolution of previous problems
- The current market

HISTORICAL DEVELOPMENT OF DESIGN–BUILD

Protagonists of Design–Build often point to historic buildings, such as Wren's St Paul's cathedral, and suggest that many of them were commissioned on a Design–Build basis, and that therefore this method of building procurement is well tried and tested. With due deference to them, and to Kipling, they could not be more wrong. The building industry has changed considerably of late, and the method of procuring buildings of old bears no resemblance whatever to present-day Design–Build in either contractual, commercial or technical terms, nor in terms of speed.

Package Deals

Design–Build, as it is understood today, started after the Second World War but failed to make any real impression until the late 1950s, when a number of building contractors began to offer the Package Deal. This was the popular name for Design–Build, reflecting the concept that contractors offered employers a complete 'package', in contrast to the comparatively fragmented traditional arrangements, whereby employers have separate agreements with designers, a professional quantity surveyor and a contractor.

Of these early Package Deal contractors, there were a minority who could be classed as design-led companies, in that their founders were designers who acquired the resources and expertise to construct buildings, and were

11

prepared to accept 'contracting' risks. However, these firms were in the minority, and this may well have been partly because the RIBA's codes of professional conduct barred architects from becoming company directors. Furthermore there may have been few architects or engineers with the desire or necessary management skills and financial backing to become involved in contracting.

The majority of Package Deal companies were existing building contractors who formed their own design departments, taking on designers as direct employees, and it was they who undertook both the architectural and engineering design work for contractors.

A number of such contractors established quite large design departments and they were successful to varying degrees. Some larger contractors acquired a high level of expertise in certain specialist fields, particularly in engineering; others developed building systems.

However, the use of Package Deal contracting did not grow to such extent that it became a major alternative form of building procurement until the recent surge in popularity dating from the early to mid-1980s.

Stagnation of Package Dealing

It is worth considering the reasons for the slow growth of Package Deal business, compared with the way that Design–Build has expanded so rapidly in recent years. There are several, and these are examined below.

The Employer's View

The number of employers who chose to follow the Package Deal route was limited for a number of reasons, as illustrated by the examples in the following list:

(i) There had been no recognised independent form of contract. The only standard form was published by the NFBTE (National Federation of Building Contractors), latterly the BEC (Building Employer's Confederation), and this could have been perceived as being biased in favour of contractors. Otherwise contractors prepared their own forms of contract, or they made amendments to the existing traditional forms, and these, obviously, could have been regarded as being even more biased.

(ii) Needless to say, any regular employer, suffering adversely from such terms, would be loath to try the same method again.

(iii)There was little acquired expertise among employers in obtaining competitive tenders for Package Deal work and so, rather than opt for non-competitive negotiations with a single contractor, they would stay with the tried and tested traditional competitive method.

(iv)Package Dealing was relatively uncommon, and it was natural for employers to show reluctance to experiment with methods of procurement which were relatively untried.

(v) Unlike today, there were few architects or engineers who were willing to accept the concept of working for, and under the direction of, a contractor, and so design work was carried out, almost exclusively, by contractors' personnel. As such they acquired the reputation of being 'jack-of-all-trades'. These individuals would be designing a warehouse one day, an office the next, and then perhaps a high-rise block of flats. These designers had little opportunity to specialise, and without depth of experience, how could they be expected, by instinct, to foresee problems particular to the building in hand, let alone to design the problems out from the start. Clearly the quality of design suffered.

The Contractor's View

Contractors, themselves, also found problems with this form of business for a number of reason, including those in the following list:

(i) Marketing and selling Package Deals was difficult. Few employers decided of their own accord to opt for the Package Deal method, and so it was up to contractors to find potential employers, and to find them at the right time. This was extremely difficult, because they had to be found before they were committed to the traditional route, but after they were prepared to divulge their plans to contractors. Having found a potential employer, it was then a new idea which contractors had to explain. Naturally, as we saw before, there was sales resistance among employers for this reason.

(ii) It generally takes a long time after a potential employer. is found, before he can give a contractor any commitment to proceed. Any scheme of size is normally subjected to an employer's internal sanctioning process, and often there are procedural and legal hurdles to overcome. Invariably proposals would be changed which added still further to the pre-contract time. Faced with this, many a contractor would give up, lose interest, lose contact through staff changes, or simply just fall out with the potential employer.

(iii)The majority of contractors' mainstream business was in Traditional Contracting. A company cannot satisfy two diverse turnover targets. So, if it meets its construction turnover target with Traditional Contract work, it would not have the capacity nor keenness to win Package Deal work simply to make up for a shortfall of work for its design department.

(iv) Experience from that time showed that for some of the reasons given, and for others, Package Deal work was invariably cyclic, and thus Package Deal companies had the problem of either maintaining uneconomic staffing levels during the troughs in work-load, or reducing the resources to suit the work in hand. The unacceptable cost of the former normally dictated that the latter choice was adopted and hence, when the demand returned, contractors had to recruit new staff or rely upon temporary staff. The effect on efficiency, morale and quality was obvious.

To sum up contractors' difficulties: Package Deal work was hard to find, and hence expensive to market and sell; it was expensive to tender over a long period, without any certainty of securing the contract in the end; and fluctuations in work with the inevitable high staff turnover was costly and undesirable.

Despite the difficulties, a significant number of Package Deals were completed satisfactorily, and a number of employers continued to use the Package Deal as a method of procurement through the 1960s and 1970s, albeit at a relatively low level, that is, no more than a few per cent of the total relevant volume of construction work.

The concept did not completely lose favour, largely because lump-sum, single-point-responsibility contracting was, and is, intrinsically attractive to employers; always provided that the framework within which it operates is satisfactory.

DESIGN–BUILD OF THE 1980S AND 1990S

Changes in the Character of the Industry

In the early 1980s, changes within and outside the construction industry were taking place. Coming out of recession, a new commercialism was being forced upon all industries; value-for-money and guarantees on time and cost were being demanded of all, as never before.

At the same time the character of construction was changing, as more and more work was being sub-let, contractors were becoming construction managers rather than the master builder and craftsmen of old.

The image and character of the contracting companies themselves was changing as more sophisticated technical, commercial, and management techniques were being employed to meet the complexities and technical advances in modern buildings.

Whilst consultants were able to stay at the forefront of design technology, they remained, as before, unwilling or unable to offer contractual guarantees on performance.

Publication of the JCT 1981 Form

These changes coincided with the publication of the JCT 1981 'With Contractor's Design' Form of Contract, and at last this was the signal that the Package Deal was a respectable and sensible way of procuring a new building. After all, the JCT had produced the Form, and the JCT consists of representatives from professional and client bodies as well as contractors. Almost to signal its 'respectability' compared with the previous arrangements, the name Package Deal was dropped almost overnight, and is rarely, if ever, used today.

The JCT 1981 Form not only provided a framework for securing competitive tenders from contractors, it also introduced the concept that an employer may engage consultants to take the design part way before the Contractor takes over to complete it.

Resolution of Previous Problems

Thus, most of the previous problems were resolved:

(i) There was an 'unbiased' form of contract.
(ii) Consultants became willing to work for contractors. This enabled contractors to use in-house designers, or external consultants, where appropriate, and hence the problem of design competence was solved.
(iii) Also, by employing outside consultants, contractors overcame the problem of fluctuating in-house design resources.
(iv) Contractors renewed their efforts in selling the concept, and employers, in increasing numbers, bought the idea, not least because the JCT 1981 Form provides a route for obtaining competitive Design–Build tenders.
(v) As the use of Design–Build expanded, so also did the understanding of its advantages and the skill in using it grew among employers.
(vi) A large proportion of Design–Build work is designed by independent professionals, and so they now energetically sell their services among contractors, and thereby, indirectly, support the system.

The Current Market

This then brings us up to date. Accurate estimates of the ratio of Design–Build work to Traditional Contracting are difficult to obtain, but estimates of 30:70 are often quoted, and that proportion is still increasing; surveys* suggest that the annual market value of Design–Build work was some £2 billion in 1989, followed by an estimated £3 billion in 1990, compared with less than £1 billion in 1984. Furthermore, the volume will no doubt increase still further, as many public sector employers are beginning to procure work by this method.

* CCMI survey, 1986; Contract Journal Survey June 1990

4 Forms of Contract

'The Form remains, the function never dies'

Wordsworth *[The River Dudden]*

KEY TOPICS

- The JCT and its 1981 form of contract
- The BPF and its system and the BPF/ACA form of contract
- The JCT 1981 BPF supplementary provisions
- The GC/WORKS/1 (Edition 3) form of contract
- Provisions of JCT 1981 and the BPF/ACA forms for:

    Contract documents (brief description)
    Design liability
    Variations
    Extensions of time

INTRODUCTION

As this book is intended, primarily, to give practical, as opposed to academic, guidance to students and potential practitioners of Design–Build, and to broaden their understanding of the way it operates in practice, the topic of contract conditions is dealt with, relatively briefly, with comment restricted to the most pertinent conditions of the contracts. Thankfully, contractual disputes in this form of contracting are relatively rare, and where a standard, unamended form of Design–Build contract has been used, few, if any, disputes have reached the courts. But this does not mean that those involved should not understand the terms and provisions of the contract.

Beware of the person who says that the contract should be signed, then consigned to the drawer. His apparent motive is to generate an atmosphere of trust and give assurance that his organisation will not manipulate the contract terms in their own interests. Whilst this may be innocently so, there has to be an understanding of the rules that cover the day-to-day dealings between the employer and the contractor, and if it is not the contract terms which govern this, what does? Furthermore, whatever any party may say about their intention to use the contract, or not, that party will, no doubt, employ legal advice and make full use of the contract terms if faced with damage or loss.

16

It is therefore, sensible for the executives of any employer, contractor, consultant or quantity surveyor to obtain an understanding of the ramifications of the standard forms of contract before their firm becomes involved in this method of contracting, and their staff should acquire a working knowledge of the terms and conditions commensurate with their particular potential involvement and level of responsibility. If faced with non-standard, or amended standard forms, specific advice should be sought from an appropriately qualified source to ensure that there are no hidden dangers, or risks, which the firm would not, otherwise, knowingly accept.

STANDARD FORMS OF DESIGN–BUILD CONTRACT

The most commonly used Design–Build form is,

'JCT 81'   Standard Form of Building Contract with Contractor's Design

The next most commonly used form is,

BPF/ACA Form   ACA Form of Building Agreement, British Property Federation Edition, 1984

A recent issue is,

JCT/BPF Form   The JCT 1981 Form, as above with Supplementary Provisions issued in February 1988, to make the form suitable for use with the BPF System

Less commonly used is,

GC/WORKS/1   General Conditions of Contract for Building and Civil
(Edition 3)   Engineering

Background to the Production of the JCT 1981 Form

The Origin of the Joint Contracts Tribunal, (JCT).

In 1931 the RIBA form of contract was agreed between the RIBA and the NFBTE and published for general use by employers and contractors. At the same time the JCT was set up by the two bodies to keep the form up-to-date. Between then and now the Tribunal, which is not in fact a 'tribunal' was increased to include representatives of various other bodies.

The bodies represented on the Joint Contracts Tribunal (JCT) are:

Royal Institute of British Architects
Building Employer's Confederation
Royal Institution of Chartered Surveyors
Association of County Councils
Association of Metropolitan Authorities
Association of District Councils

Confederation of Associations of Specialist Engineering Contractors
Federation of Associations of Specialists and Sub-contractors
Association of Consulting Engineers
British Property Federation
Scottish Building Contract Committee

During its existence, the JCT has published a considerable number of contract forms and revised editions to suit the changing needs of the industry, and to accommodate changes in statute where applicable. The following major editions have been issued by the JCT since 1931:

'Traditional' contracts
  The 1931 'RIBA' Form
  The 1939 Revised Edition
  The 1963 Edition
  The 1980 Edition

Since 1980 the JCT has issued standard forms for non-traditional contracts, including the 1981 'With Contractor's Design' form and the 1987 Standard Form of Management Contract.

The JCT has also published sub-contract forms, and practice notes for use with all the current forms.

The Origin of the JCT 1981 Form

It took the JCT from 1977, when a request was made by the Department of the Environment to the Tribunal for the production of a standard form for Design–Build contracting, to 1981 to devise, agree and issue it; there was no forerunner which divided the design responsibility between the employer and the contractor, for the JCT to use as a basis in drafting the form.

Background to the BPF System and BPF/ACA Form

The British Property Federation is represented by its general council, on which sit some twenty-four members. They had become increasingly concerned about the problems occurring, in their opinion, far too often in the design and construction of buildings. As a result, they formed a working party of seven members with a technical sub-committee and consultants to look at ways to improve upon the commonly used systems of procuring both the design and the construction of buildings.

The Manual of the BPF System

In 1983/4 the BPF published its findings and recommendations in the form of the Manual of the BPF System.

The System covers the complete process of the design and construction of a building, and deals with it in five stages, namely:

(i) Concept
(ii) Brief
(iii)Design development
(iv) Tender documents and tendering
(v) Contractor's design and construction

The Manual describes the role of each party at each of the defined stages with the object of removing duplication between parties and avoiding split responsibilities; thereby, it suggests, the client will obtain a good building more quickly and at a lower cost.

Pre-contract Design – BPF System

As regards the design of the building, the System primarily envisages that designers employed directly by the client shall design the building almost in its entirety, leaving the contractor to design those parts which are most appropriate for him to design. For example, where the method of construction is an important factor in the design or detailing, as is the case with foundations and structural work.

Forms of Contract for Use with the BPF System

With the publication of the BPF System, the BPF/ACA Form of Building Agreement was published, and subsequently, in 1988, the Supplementary Provisions to the JCT 1981 Form were published for use with the BPF System.

Main Provisions of the JCT/BPF Supplement

The provisions of the supplement are numbered S1 to S7 covering topics as follows:

S1 – Adjudication
  Certain specified matters of dispute may be referred by the employer or the contractor to an adjudicator for speedy settlement.
S2 – Submission of drawings, etc. to the employer
  The contractor shall submit all drawings, etc. to the employer for comment prior to implementation.
S3 – Site manager
  The duties of the site manager are specified more fully.
S4 – Named subcontractor
  The employer may 'name' subcontractors in the Employer's Requirements, and this provisions lays down the conditions that apply.
S5 – Bills of quantities
  Bills of quantities may be used.
S6 & S7 – Variations and loss and expense
  The rules for valuing variations are similar to the BPF/ACA form.

Background to the GC/WORKS/1 (Edition 3) Form

The GC/WORKS series of contracts are issued by the Department of the Environment. Unlike its predecessors in the series, Edition 3 contains a clause which provides for design of part of the works to be undertaken by the contractor; (Clause 10. (1)–(4)).

It would appear to be the intention of the authors, that the contractor would only be required to carry out the design of discrete parts of the works, and in no sense to take responsibility for the design or function of the works as a whole. In this respect, it is similar in intent to the BPF System.

However, the GC/WORKS/1 Edition 3 form is rarely used, in practice, for Design–Build projects.

PROVISIONS OF THE STANDARD FORMS

As it is the objective of this book to give an account of the practical use of Design–Build, we shall only look at those aspects in the JCT 1981 and the BPF/ACA forms which are particular to Design–Build, as opposed to Traditional Contracting, and those which are also of most relevance to Design–Build practitioners on a general day-to-day basis.

These topics are:

(i)   Contract documents
(ii) Design liabilhty
(iii) Variations
(iv) Extensions of time

Contract Documents, JCT 1981 Form

The contract documents required for the JCT 1981 contract are:
(i)   Employer's Requirements
(ii) Contractor's Proposals
(iii) Contract Sum Analysis

Recommendations on the contents and preparation of these contract documents are contained in the JCT'S Practice Notes CD/1A and CD/1B.

The Employer's Requirements, JCT 81 Contract Document

The Employer's Requirements is a statement of the employer's requirements and it can be exceedingly brief or on the other hand it can be lengthy and detailed. In either case it is the employer's responsibility to have it prepared.

It is important to differentiate between this document and the requirements of the employer as expressed in the enquiry documents. Whilst the employer may set out to prepare the enquiry in such a way that it can be incorporated into the contract documents as it stands, in practice it is exceedingly rare to be able to do so without some amendments.

These amendments may be such as to constitute a complete re-draft of the requirements, or they may require little change or editing. Often letters or other post-tender documents are incorporated, so that the Employer's Requirements becomes in essence the 'enquiry' plus a set of other documentation.

One addition which is often necessary, is the incorporation of provisional sums. The JCT 1981 form has no terms which deal with provisional sums, unless they are specifically included within the Employer's Requirements. So, if the tender which is accepted, contains provisional sums, and both parties accept them, then the Employer's Requirements must not only include them, but it must also include the method of valuing works which are the subject of the provisional sums.

Contractor's Proposals, JCT 81 Contract Document

It is unlikely that a contractor's tender has ever been accepted and incorporated into the Contract Documents without an amendment of some kind. Hence, we must differentiate between the tender, which the contractor may, nevertheless, have called the contractor's proposals, and the Contractor's Proposals which do, in due course, form part of the contract.

The Contractor's Proposals complement the Employer's Requirements, and should be checked to ensure that no discrepancy exists between the two documents. Together the documents should contain sufficient drawings and specification details to describe the precise nature of every element of the proposed project, or where this is impractical, the documents should define the standards or parameters that are to apply in the subsequent design development.

In other words, the Contract Documents should be so prepared that no one gets any surprises when the details are developed and the works constructed. This is in the interests of both the employer and the contractor. For the employer the benefits are obvious; for the contractor it may seem to be helpful to have a vague specification that will enable him 'to get away with' sub-standard products, but in practice he will usually lose out where the work is not properly defined.

For example, *a contractor may price for a medium quality component, such as ironmongery, and then, if the specification is unclear, offer a lesser quality. The employer may be in a position to demonstrate that the component is not of a suitable standard and demand better, and perhaps even better than the contractor had allowed in his tender; without the detailed specification the contractor would have difficulty in winning the argument. Not only that, if the contractor incorporated the lesser quality component within the works, it may fail at a later date, and he is then faced with replacement costs.*

Thus, to sum up, it is in both parties interests to take the time and trouble to ensure that the Employer's Requirements and the Contractor's Proposals do, together, define the Works specifically, and in detail.

Precedence – Employer's Requirements v Contractor's Proposals

It is not as simple as it may first appear to pre-determine the rules for precedence in the event of a discrepancy being discovered between the Employer's Requirements and the Contractor's Proposals.

The test to determine precedence must take in account the following provisions:

(i) The third Recital of the standard JCT 1981 Form says that the Employer has examined the Contractor's Proposals and that he is satisfied that they *appear* to meet the Employer's Requirements.

(ii) Clause 2.5.1, in effect, obliges the Contractor to use reasonable skill and care in the design of the Works comprised in the Contractor's Proposals.

Thus it can be seen that if a contractor does not use reasonable skill and care, and for this reason a discrepancy arises, then the contractor is at fault, and he must therefore remedy any problem arising without recompense.

It may be deduced from the third recital that the Contractor's Proposals will take precedence in discrepant items which are obvious and fundamental to the Contractor's Proposals.

For example, *if there were discrepancies between internal finishes, described in the Employer's Requirements, and those clearly described within a finishes schedule contained within the Contractor's Proposals, then it would be hard to conclude other than that the Contractor's Proposals should take precedence.*

On the other hand, if the discrepancy is more subtle, the conclusion would be less easy to draw.

For example, *if the Employer's Requirements stipulated that all lighting levels were to be in accordance with CIBSE guidelines, and the Contractor's Proposals contained a schedule of specific lighting levels, which, in some instances, were contrary to the guidelines, would the Contractor have used reasonable skill and care in his design? If not, then he would be in breach of his obligations and therefore liable to remedy the situation.*

Contract Sum Analysis, JCT 1981 Contract Document

As there is no bill of quantities in this form of contracting, the JCT have introduced, instead, a Contract Sum Analysis. Its purpose is to assist in the valuation of changes in the Employer's Requirements and for valuing work which is the subject of provisional sums. It is also normally used as the basis for interim valuations.

BPF/ACA Contract Documents

The documents required for the BPF/ACA Contract are as follows:

- (i) Contract Drawings and Specification
- (ii) Contract Bills, if used
- (iii) Time Schedule
- (iv) Schedule of Activities

BPF/ACA Contract Drawings and Specification

It is the intention of the BPF System that the Contract Drawings and Specification are prepared by the Client or his consultants. There is no provision for 'Contractor's Proposals' (cf. the JCT 1981 form) although, in practice, a contractor may be appointed by way of a letter of intent in the pre-contract period and, by agreement, prepare some of the drawings and specification for the contract documents.

In whatever way the drawings and specification are prepared they should fully describe the Works, or lay down the parameters within which the Contractor shall design the remaining parts of the works.

BPF/ACA Contract Bills

The BPF System is so drafted that a bill of quantities can be used for the tendering and valuation of part, or all, of the Works. Having said this, it is the stated intention of the authors of the System that bills should not normally be used, and that the lump-sum concept should apply.

BPF/ACA Time Schedule

The Time Schedule consists of the following two parts, neither of which is a programme, as its name may imply.

- (i) date(s) for possession and completion
- (ii) schedule of drawings and details, etc. which the Contractor is to provide to the Client's Representative (for sanction) giving planned dates for the issue of each.

The Client will normally specify the dates in (i), and the Contractor will produce the schedule in (ii).

BPF/ACA Schedule of Activities

The BPF Manual gives guidance on the make-up of the Schedule of Activities, suggesting that the following information should be given for each activity:

(i) Activity description
(ii) Quantity
(iii)Resources
(iv)Start time (week number)
(v) Duration
(vi)Price

The Schedule of Activities is to be prepared by the Contractor. The BPF Manual gives typical examples of the schedule; in an abbreviated form to be submitted with the tender and in a developed form for use during the contract. It distinguishes between the treatment of 'measured', or measurable, work and 'non-measured' elements such as preliminary costs.

To encourage contractors to complete their work in a regular and pre-determined way, the BPF/ACA contract provides for interim payments to be based only upon *completed* activities as defined and described in the schedule.

DESIGN LIABILITY

Design Liability, in General

In Chapter 11 we shall see the full extent of the risks and responsibilities borne by contractors in a Design–Build project. They include risks of a commercial nature. Here we briefly consider the nature and extent of a contractor's duty, or legal obligation, to the employer to design the building satisfactorily and without defects.

The duty is defined differently in the two standard Design–Build contracts which we are considering, and indeed in the many amendments that employer's tend to make to the standard forms.

However, the contractual liability will, in prinbiple, fall within one of the following categories:

(i) Duty to exercise reasonable skill, care and diligence in the performance of the design work, or
(ii) Duty to design the works, or parts of the works, such that they will be fit for their intended purpose(s).

There is a fundamental difference between these two levels of duty.

Duty to Exercise Reasonable Skill, Care and Diligence

In discussing a contractor's duty to the employer, it is relevant to consider an architect's, or engineer's duty to their client, because the contractor's liability in the JCT 1981 Form is described as being similar to that of a professional designer.

The law hardly differentiates between professions in determining the duty or obligation which a 'professional' owes to his employer or client. So, the duty of accountants, architects, barristers, dentists, doctors, professional engineers, professional quantity surveyors, solicitors, etc. may all be similarly described; and that is:

> *When a professional person agrees to provide his professional services to a client, and no contract exists with terms to the contrary, the professional owes a duty to the client to exercise reasonable, skill and care in conformity with the normal standards of his profession.*

This does not mean that he is guaranteeing to achieve a desired result. For example, a doctor does not guarantee to cure a patient, even if he is curable; a defending barrister does not guarantee acquital for his innocent client. They do not even guarantee to perform to the standard of an average practitioner of their profession.

Unless the professional has entered into a more onerous contract, he is only legally liable when things go wrong, not just if the standard of his service was below the average which may be expected from the profession, but is so far below that standard that it could be regarded as being 'abnormal' or negligent.

Similarly, unless an architect or engineer agrees in contract otherwise, their obligation is to use reasonable skill and care in designing a building, but they do not give an absolute guarantee that the building will be free of defects; to sue them successfully in the event of a defect becoming apparent, it would be necessary, in effect, to prove negligence.

Furthermore, the professional indemnity insurance held by consultants, with probably few exceptions, is restricted to claims for damage as a result of negligence, and not as a result of a more onerous contractual liability, as for example is covered in the next paragraph.

Fit-for-Purpose Obligation

A fit-for-purpose obligation, or condition, means that the provider of the service is obliged, absolutely, to achieve the required result. This obligation contrasts with the professional duty described above, in that it is a 'strict' liability, or a 'no-fault' liability. Thus, it is a far more onerous obligation.

In designing a building, a designer who accepts an obligation to design the building, so that it will be fit for the purpose for which it is intended, will be liable for any design deficiency which subsequently renders the building, or part of it, unfit for its intended purpose; it is immaterial whether the designer in question, or any other designer, could have possibly foreseen that the design would be deficient in any way.

In other words, to sue such a derigner successfully, it would not be necessary to prove negligence, but merely to prove that the building, or part, was not fit for its intended purpose. The question of what is the purpose of the building can only be answered in each specific case, but from this it can be seen that any contractor, or designer, should ensure that his understanding of the purpose of a building, and its parts, is clear and agreed with the employer.

Uncertainty of the Law in Respect of Design Liability

Generally the liability which a designer has to his client will be specified in writing either in an agreement or in a letter, which may refer to standard terms of appointment published by the designer's professional institution. Where there is no written agreement at all, as for instance there may be if the two parties failed to confirm any agreement in writing, or in the case of a third party, e.g. a subsequent purchaser of the building, the situation with regard to liability is ever changing, in accordance with precedence created by successive court decisions.

Similarly, the situation with regard to recoverable losses is by no means clear.

The 1985 NEDO report, Latent Defects in Buildings: an Analysis of Insurance Possibilities, includes the following statement:

'The English Law on construction liability is a mess. It is incomprehensible to a layman and apparently also to lawyers who specialise in it. Contractors, professionals and suppliers are not sure of their obligations and rights, and building clients are not sure when they have a valid claim'.

Notwithstanding this quotation, these are areas where the layman should seek the advice of lawyers. The following, but not exhaustive, lists of statutes and cases, which could be relevant in determining the outcome of a dispute, serve to illustrate that the difficulty that one may have in predicting a court's decision in any new case.

Statutes which may be relevant:

(i)   Sale of Goods Act, (1979)
(ii)   Unfair Contract Terms Act, (1977)
(iii) Defective Premises Act (1972)
(iv)   Building Act (1984)

Court decisions which may be relevant:

(i)   Bolam v Frien Hospital Management Committee (1957)
(ii)   Bagot v Stevens Scanlon (1963)
(iii) Corben v Hayes (1964)
(iv)   Moresk Cleaners Ltd. v Thomas Henwood Hicks (1966)
(v)   Young and Marten Ltd. v McManus Childs (1968)
(vi)   Greaves and Co (Cont'rs) Ltd. v Baynham Meikle & Pts. (1975)
(vii) Midland Bank Trust Co. v Hett Stubbs & Kemp (1978)
(viii) Anns v London Borough of Merton (1978)

(ix)   Batty v Metropolitan Property Realisations Ltd. (1978)
(x)   IBA v EMI Electronics and BICC Construction Ltd. (1980)
(xi)   Investors in Industry v South Beds. District Co. (1980)
(xii) London B'gh of Newham v Taylor Woodrow-Anglian Ltd. (1981)
(xiii) Pirelli v Oscar Faber & Partners (1983)
(xiv) Sunnyside Nursing Home v Builder's Contract Management Ltd. (1985)
(xv)   Basildon District Co. v J E Lesser (Properties) Ltd. (1985)
(xvi) Viking Grain Store Ltd. v T H White Installations Ltd. (1985)
(xvii) Junior Books v Veitchi (1985)
(xviii)Consultant's Group International v John Worman Ltd. (1985)
(xix) Peabody Estates v Sir Lindsay Parkinson Ltd. (1985)
(xx)   Tai Hing Cotton Mill v Liu Chong Hing Bank (1986)
(xxi) Yuen Kun Yeu v A G of Hong Kong (1987)
(xxii) Greater Notts. Co-op. v Cementation (1988)
(xxiii)Department of the Environment v Thos. Bates & Son (1987)
(xxiv) D & F Estates v Church Commissioners (1988)
(xxv) Pacific Associates v Baxter (1988)
(xxvi) Portsea v Michael Brashier & Associates (1989)

As the length of the list implies – and the list is not exhaustive – this is a complex topic which cannot be covered in any depth here. For the purpose of this book, it is, perhaps, sufficient to note the main points, as follows:

(i)   Unless there are specific terms to the contrary in an agreement with a designer, his obligation will be restricted to the exercise of reasonable skill and care.
(ii) In complete contrast, a contractor (a supplier of goods and services) will, in all probability, have an absolute 'fitness-for-purpose' liability, unless the contract terms specifically exclude or limit his liability.
(iii) Some exclusion clauses may be disregarded by the courts, by reason of the Sale of Goods Act or the Unfair Contract Terms Act.

Contractor's Design Liability, JCT 1981 Form

Article 1 of the the JCT 1981 Agreement says that the Contractor shall complete the design of the Works. He will have also prepared the Contractor's Proposals forming part of the Contract. Clauses 2.5.1, .2, .3, and .4, of the Contract, stipulate the Contractor's design warranty and liability in respect of both these design stages

Clause 2.5.1 is a detailed statement of the Contractor's liability, and it says, in essence, that the Contractor shall have the like liability of a professional designer in respect of any defect or deficiency. As we have seen this means that the contractor-designer is required to exercise reasonable skill and care in the design work, in conformity with the normal standards of the relevant profession.

Clause 2.5.2 deals with the additional liability in respect of compliance with the Defective Premises Act 1972, if applicable to the particular project.

Clause 2.5.3 provides for a limit to be put upon the amount of the Contractor's liability to the Employer for losses due to defective or deficient design.

Clause 2.5.4 lays down that there is no difference, in respect of the Contractor's design liability, between design undertaken by the Contractor himself and design which he may procure from others

Contractor's Design Liability (BPF/ACA Form)

We have already seen that in the BPF System, the client generally completes a considerable proportion of the design of the project, leaving the contractor only to design parts of the project, and it is in this light that we consider the contractor's design liability,

In essence, the contractor's design obligations are described as follows:

Article F of the Agreement obliges the Contractor to provide all further drawings and information etc. for the execution of the Works in accordance with the provisions of Clause 2.2 which says, in effect, that:

The Contractor shall submit, for sanction, drawings and information etc. which are

(i) needed to amplify the Contract Drawings and Specification
(ii) needed by the Contractor to execute the Works
(iii) stated within the Contract Documents, to be provided by the Contractor.

Clause 3.1 says that the Contractor warrants that the Works will comply with any performance requirements contained in the Contract Documents, and that those parts of the Works to be designed by the Contractor will be fit for the purpose for which they are required.

Note that both these requirements are absolute, and not merely an expression that the Contractor shall only be required to exercise due skill and care.

Clause 5.4 requires the Contractor to employ appropriately skilled persons as may be necessary to discharge the Contractor's obligations.

Clause 6.6 requires the Contractor to take out and maintain insurance cover in respect of negligence in his design of the Works.

Thus, we see that with the BPF/ACA Form, contractors have a strict no-fault liability with respect to those parts of the Works which they are to design, but they have no responsibility for the design implicit within the contract documents. They have no responsibility for the functioning of the building as a whole. The design indemnity insurance that they are required to take out does not cover this no-fault liability, but refers only to problems of negligence.

VARIATIONS

In the context of this chapter, variations are any changes in design, circumstances, or requirements, etc. which lead to a change in the contract sum, in accordance with the terms of the contract.

Variations – JCT 1981 Form

Clause 13 of the JCT 1981 Form states, in effect, that the Contract Sum shall not be adjusted or altered, except in accordance with the provisions of the Contract. It does not summarise the relevant clauses, which would have made the interpretation of the clause more convenient. However, the following lists include all the items which, in accordance with the Contract, may lead to a change in the Contract Sum:

Changes Within the Employer's Control

(i) Employer's instructions given to correct errors in the definition of site boundaries in the Employer's Requirements; Clause 2.3.1.

(ii) Costs of inspections or tests instructed by the Employer, where such tests show the work to be satisfactory; Clause 8.3.

(iii) Payment of royalties, under certain circumstances, when the need arises through an Employer's instruction; Clause 9.2.

(iv) Changes in the Employer's Requirements; Clause 12.2.

(v) Expenditure of provisional sums contained within the Employer's Requirements; Clause 12.3.

(vi) The cost of making good defects during the Defects Liability Period, which are not the fault of the Contractor; Clause 16.3.

(vii) Under the circumstances described, where the Employer insures the Works under option 22C of the Contract, the costs of reinstatement and removal of debris after loss or damage to the Works; Clause 22C.4.4.2.

(viii) Insurance premiums, where the Employer is in default of his obligation to pay such premiums; Clause 22B.2 or 22C.3.

(ix) Losses caused through disturbance to progress caused by others employed directly by the Employer; Clause 26.2.3.1.

(x) Losses caused through disturbance to progress caused by delay in receipt of materials which the Employer is contracted to supply; Clause 26.2.3.2.

(xi) Employer's instructions with regard to postponement of the work; Clause 26.2.4.

(xii) Losses caused through the Employer's failure to give access or egress to the Contractor in accordance with the conditions of the Contract; Clause 26.2.5.

(xiii)Losses arising from delay caused by the Employer not giving instructions or information to the Contractor in due time; Clause 26.2.7.

Changes Beyond the Direct Control of the Employer

(i) Statutory fees or charges, where they are expressed as provisional sums within the Employer's Requirements; Clause 6.2.
(ii) Costs arising from changes in statutory requirements made after the Base Date named in the Contract; Clause 6.3.
(iii)Losses caused through delay in receipt of development control approvals which the Contractor had taken all practicable steps to avoid; Clause 26.2.2.
(iv) Losses in connection with war or hostilities; Clauses 32 & 33.
(v) Losses arising from the discovery of antiquities; Clause 34.
(vi) Fluctuations, if applicable; Clauses 36–38.

Changes, Deductions, Due to the Contractor's Default

(i) Cost work by others as a result of default by the Contractor; Clause 4.1.2.
(ii) Breach by the Contractor of provisions relating to insurance; Clauses 21.1.3, 21.2.3 and 22A.2.
(iii)Liquidated damages for delay; Clause 24.2.1. Although it must be added that, strictly speaking, this does not constitute a change in the Contract Sum as, in fact, it is a separate payment or debt owed to the Employer by the Contractor.

Variations (BPF/ACA Form)

Clause 15 of the BPF/ACA Form conveniently lists those clauses (and no others) which may give rise to an adjustment in the Contract Sum. They are described in the following lists:

Changes Within the Client's Control

(i) If contract bills are used in the Contract, i.e. if alternative 2 to Clause 1.4 is used, the Contract Sum is to be adjusted to correct any mistakes in the bills.
(ii) Disturbance to the regular progress of the work caused by any act, omission, default or negligence of the Client or his Representative; Clause 7. Note that the acts, etc., to which this clause applies, are not limited to those where the Client is necessarily at fault.
(iii)Client's instructions to accelerate or postpone the works; Clause 11.8.
(iv) Remedial works carried out during the Maintenance Period, upon the instructions of the Employer, which have not been caused by the Contractor; Clause 12.3.
(v) Client's Representative's instructions; Clause 17.

Changes Beyond the Direct Control of the Client

(i) Compliance with statutory requirements, in certain circumstances only; Clause 1.7.
(ii) Payments related to occurrences covered by the insurance of the Works; Clause 6.4.
(iii) Fluctuations, if applicable; Clause 18.

Changes, Deductions Due to the Contractor's Default

(i) Failure on the part of the Contractor to insure the Works in accordance with his obligations under various clauses of the Contract; Clause 6.9.
(ii) Failure on the part of the Contractor to make justifiable claims upon the contract insurances which he was obliged to take out; Clause 6.10.
(iii) Breach on the part of the Contractor in respect of defective work; Clause 12.4.
(iv) The Employer may deduct damages, for delay in completion, under Clause 11.3, but this is not mentioned in Clause 15 as this is not regarded as a change in the Contract Sum, but as a debt owing to the Employer.

Clause 15 also includes Clause 25.2, the adjudication clause, among those whose provisions may give rise to a change in the Contract Sum.

EXTENSIONS OF TIME

The grounds for extension of the contract period vary little between the JCT 1981 and the BPF/ACA forms of contract, and so we shall consider the two forms side-by-side. Contractors should beware of deletions which employers are accustomed to make. Employers on the other hand should recognise that the more risk to the contractor in the contract, the more contingency, or greater profit, he will allow. These extra costs become compounded because, not only do the subcontractors weigh up the risk and include appropriate contingencies, but the contractor compounds it by adding a percentage to the costs, in addition to including his own contingency. It may even tip the balance and cause one or all the tendering contractors to decline to tender, or to tender without any sense of keenness, which, of course, would be ultimately to the employer's detriment and cost.

It must be remembered that the granting of an extension of time by an employer, does not automatically entitle the contractor to recovery of any loss and expense arising from the delay; such losses are only recoverable under the relevant provisions for the adjustment of the contract sum, dealt with earlier in this chapter.

Table 4.1 shows the grounds for extension of time included within the unamended standard JCT 1981 and the BPF/ACA forms of contract.

Table 4.1   Grounds for Extension of Time, JCT 81 and BPF/ACA Contracts

EVENTUALITY	JCT1981 clause	BPF/ACA clause
Force majeure	25.4.1	11.5(a)
Exceptionally adverse weather	25.4.2	N/A
Damage in connection with insurable matters	25.4.3	11.5.(b)
Civil commotion, strikes, etc.	25.4.4	see below
War, civil commotion, etc., not strikes	see above	11.5.(c)
Compliance with employer's instructions	25.4.5.1	11.5.(e)
Opening-up for inspection, proved needless	25.4.5.2	11.5.(e)
Late receipt of employer's instructions, etc	25.4.6	11.5.(e)
Delays through late statutory approvals, etc.	25.4.7	N/A
Delay by employer's other contractors	25.4.8.1	11.5.(e)
Delay in receipt of employer's materials	25.4.8.2	11.5.(e)
Statute changes affecting labour availability	25.4.9	11.5.(d)
Inability to obtain labour	25.4.10.1	N/A
Inability to obtain materials	25.4.10.2	N/A
Delays by LA or stat. undertakers works	25.4.11	11.5.(d)
Failure by the employer to give access	25.4.12	11.5.(e)
Delay due to changes in stat. requirements	25.4.13	11.5.(d)
Deferment of possession of site	25.4.14	11.5.(e)

General Rules Applying to Extensions of Time

The descriptions of the grounds for extension, given in Table 4.1, are abbreviated considerably and it should be noted that many of the eventualities would not lead to an extension of time unless a number of pre-conditions are met by the contractor. To understand the full implications it is necessary to study the contract terms themselves in more detail.

However, as a general rule, a contractor will only secure an extension of time if he can show that he had taken all reasonable steps in order to prevent or minimise the delay; that he had given the proper notices to the employer; and that he could not have reasonably foreseen the eventuality at the time the Agreement.

Furthermore, the contractor will also have to demonstrate that any delay to the work in progress inevitably leads to a delay in the completion date. Also, any delay or extension of time relating to a sectional completion date would not necessarily lead to an extension of time for other contractual completion dates.

Both contracts make provision for an extension of time, granted by an employer, to be challenged by the contractor by seeking the opinion of an arbitrator (or adjudicator in the case of the BPF/ACA).

5 Varieties of Design–Build Procurement

'That's our system, Nickleby.
What do you think of it?'

Dickens   [Nicholas Nickleby]

KEY TOPICS

- Different varieties of Design–Build procurement.
- Varying relationship between employer's and contractor's design input.
- Typical contents of enquiries for:

  Employer-led-design;
     Develop and Construct
     Design–Build (single-stage tender)
  Contractor-led-design;
     Design–Build (two-stage tender)
     Negotiated Design–Build
     Design and Manage
     Turnkey

- Examples of circumstance appropriate to each variety.

INTRODUCTION

An employer who wishes to use Design–Build for his project must prepare an enquiry in some form or other. This is a simple statement of fact, but deciding upon the content of a Design–Build enquiry is far from simple. The employer (with his advisors, as necessary) must not only make a number of decisions which relate to the form, terms and conditions of the contract, but he must also decide upon the extent of the technical contents of the enquiry.

Even in Traditional Contracting, where in principle employers' consultants are responsible for the full design, detailing and preparation of working drawings, there is often confusion as to how much 'design' the consultants will actually prepare themselves, and how much will be left to specialist subcontractors or suppliers. Often this can lead to disputes, both between an employer and his consultants, especially in the matter of fees; and between the consultants and the main contractor over the question of responsibility for the design and coordination, and delay, if any, in the issue of subcontractors' or suppliers' information.

Once a contractor is appointed in a Design–Build situation, there should be no such confusion; because the essence of this form of contracting is that the contractor is responsible for the production of all the information which will be required to complete the design and detailing of the works.

However, the question remains; how much design work should an employer undertake before engaging a contractor, or in other words, how detailed should the enquiry be?

To an inexperienced employer, or inexperienced advisors, this question often poses more difficulties than any other.

On the one hand, an employer and his advisors may be tempted to 'go too far' with the design, often because they find it difficult to adjust to the idea that a contractor may be capable of conceiving the correct or the best design solutions; and often because they feel that without detailed instructions in the enquiry, they will not be able to compare tenders on a like-for-like basis. By going too far with the design, an employer may incur unnecessary fees, and deny the contractor the opportunity to contribute innovatively to the design process.

On the other hand, if the employer provides too little information in the enquiry, the tenders he obtains may not include any satisfactory design solutions, in which case real time will been have lost and, in consequence possibly, the opportunity to secure value through competition. The opportunity to get the best from the tendering contractors could also have been lost, and the tendering contractors would have been put to much needless expense.

As we shall see, there is no single answer to the question. Every job is different, the circumstances which prevail at the time are unique, and each employer has his own past experiences and preferences which will influence his approach. In this chapter we refer briefly to the circumstances which may influence an employer's choice, although the topic is covered more thoroughly in Chapter 6, 'Which Form of Procurement?'.

THE DIFFERENT VARIETIES OF DESIGN–BUILD

Within the overall concept of Design–Build, a number of names have emerged to describe what we might call different 'varieties' of Design–Build procurement. They do not necessarily relate to the terms and conditions of the subsequent contract. The difference between them can usually be related to the proportion of the design, undertaken by the employer's consultants, which is included in the enquiry.

Figure 5.1 shows the six most commonly accepted varieties of Design–Build and is a useful illustration of the difference between each.

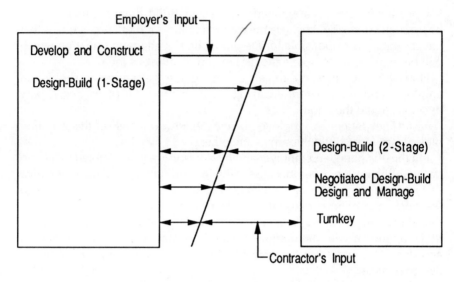

Figure 5.1 The relationship between the employer's input and the contractor's input for the different varieties of Design–Build. (The arrowed lines are indicative and not exactly to scale.)

The six varieties are:

(i)   Develop and Construct
(ii) Design–Build (single-stage tender)
(iii) Design–Build (two-stage tender)
(iv) Negotiated Design–Build
(v)   Design and Manage
(vi) Turnkey

Whilst Figure 5.1 shows a gradual shift of responsibility for the design input from employer to contractor through the six varieties, there is a marked change in emphasis after the first two varieties on the list. In these, the employer's input is such that it is necessary for him to have a relatively full design team for the production of the pre-enquiry design; whereas, in the remaining varieties, the employer is involved in relatively little design, and it is the contractor who is responsible for the principal part of the design.

Hence, we might distinguish between them as follows:

Employer-led-design	**Contractor-led-design**
Develop and Construct	Design–Build (2-stage tender)
Design–Build(1-stage tender)	Negotiated Design–Build
	Design and Manage
	Turnkey

DEVELOP AND CONSTRUCT

Develop and Construct is shorthand for 'develop the detail from the employer's design, and construct the works'.

With a Develop and Construct enquiry, employers have, or engage, their own design team. They will design the building almost completely, and invariably obtain detailed planning consent.

Contents of a Typical Develop and Construct Enquiry

The following list shows the contents of a typical Develop and Construct enquiry:

DRAWINGS (typical Develop and Construct enquiry):

Architectural	Plans of all floors and roof	1:100
	All elevations	1:100, 1:50
	Principal sections	1:50
	Some detailed plans, e.g. toilets	1:20
	Typical details of junctions, etc.	1:20, 1:5
	Details of particular features	1:20, 1:5
	External works	1:200, 1:100
Structural	Foundation, general arrangement	1:100
	Structure, general arrangement	1:100
	Sundry details	1:20
	External works construction	Various
Mechanical	Radiator/heater layouts	1:100
services	Main pipe routes	1:100
	H & C water supply routes	1:100
	Tank sizes and locations	1:100
	Boiler/calorifier layout	NTS
	A/C and vent plant layout	NTS
	Duct runs	1:100
	Schematics, all systems	NTS
Electrical	Distribution diagrams	NTS
services	Floor layouts showing:	1:100
	lighting	
	small power	
	fire alarms	
	emergency lighting	
	public address	
	telephone, TV, etc.	
	distribution boards	

SPECIFICATIONS (typical Develop and Construct enquiry):

Particular	Foundations
specifications	Superstructure
	External walling
	Windows and curtain walling
	Cladding
	Roofing
	Rainwater disposal systems
	External doors
	Internal walls
	Floor, wall and ceiling finishes
	Joinery
	M & E installations
	Sanitary ware
	Lift, hoist, escalator, etc. installations
	Hard and soft landscaping
	Drainage
	Incoming services

General specification for workmanship and materials

for all trades.

The particular specifications will be, normally, part by performance, and part by product name or description.

The specifications for the M & E installations, in particular, will usually be by performance.

Generally contractors are left to carry out the structural design, detailing and scheduling, although this could be by a named specialist subcontractor for some elements of the work, as follows:

Often, in a Develop and Construct project, in the pre-enquiry stage, the employer's consultants will obtain quotations, and designs, from specialist subcontractors, and the enquiry will then stipulate that the successful contractor will take on these subcontractors. They will not be 'nominated' in the Traditional Contracting sense, but they will be 'named' and engaged by the contractor as 'domestic subcontractors'.

The trades most often dealt with, in this way include:

Piling, structural steel or concrete structures
Curtain walling
Building services

Even though employers take the design to an advanced state in Develop and Construct projects and obtain detailed Planning Permission, it is normal for the contractor to seek Building Regulation approval.

Circumstances in which Develop and Construct may be Used

It will be noted that the heading refers to 'circumstances' and not to types of buildings themselves. This is because the choice of the variety of Design–Build procurement which may be used, in any particular case, will invariably depend on the prevailing circumstances, and only rarely on the form of the building itself.

The following list shows some examples where Develop and Construct could be used to good effect:

(i) When an employer decides to adopt a Design–Build approach even though he has the design already substantially complete, e.g. when,
   (a) he had previously commissioned the design, intending to adopt the Traditional Contracting approach but then subsequently changes his mind, or
   (b) the employer, for example, a property developer, purchases a site from another who had previously commissioned the design for use with Traditional Contracting, and the new owner prefers Design–Build.
(ii) When an employer requires a series of new buildings to be built on a repetitious basis to a standard design, for example:
   hotels, chain stores, houses, depots, etc.
(iii)When the employer, himself, is a designer.

There are many other examples where Develop and Construct is likely to be used by an employer, and a method which could assist in determining which option to choose is described in Chapter 6 'Which Form of Procurement?'.

DESIGN–BUILD (SINGLE-STAGE TENDER)

The dividing line between Design–Build (single-stage tender) and Develop and Construct is indistinct. They both come within the employer-led-design category, in that the employer's consultants carry out the design which dictates the general appearance, form and performance characteristics of the building, but this variety leaves more for the contractor to design.

With Design–Build (single-stage tender), the employer has two objectives which are, to some extent, mutually incompatible. On the one hand, he aims for like-for-like tenders, as he would receive from contractors in response to a Develop and Construct enquiry; on the other hand, he recognises the benefits of obtaining new ideas from the tendering con-tractors, and so wishes to give them as much scope as possible in their proposals.

To satisfy the first objective, the employer's consultants will carry out the conceptual design of the building. It will be in sufficient detail to identify the layout and the elevational treatment of the building so that detailed Planning Permission could be sought. It will specify the major materials by either type or performance, and it will specify the performance requirements of the building service installations.

The employer's second objective will be satisfied by leaving as much as possible to the contractor to design and specify within the framework of the conceptual design.

To respond to such an enquiry, a contractor will need to appoint his own design team (in-house, or externally) to give him advice at the tender stage, and then subsequently if he wins the contract, to fulfil the post-contract design responsibilities. These may include, among others, the resolution of Reserved Matters attached to a Planning Approval previously obtained by the employer, and the securing of approval under Building Regulations.

Contents of a Typical Design–Build (Single-Stage Tender) Enquiry

The following list shows the contents of a typical Design–Build (single-stage tender) enquiry:

DRAWINGS

Architectural	Floor plans, often as line diagrams without showing wall thickness, etc.	1:100, 1:200
	Elevations	1:100
	Sections, as line diagrams	1:50, 1:100
	External works layouts	1:200
Structural	Grid lines and column positions	1:100
Building services	Probably no drawings, refer to the specification	

SPECIFICATIONS

Particular specifications	Materials on which Planning Approval depends	by name
	Any other preferred materials	by name
	Building services	by performance
	Acoustics, if relevant	by performance

All other specifications are optional.

STANDARDS

Often, employers like to include in their enquiries an all embracing statement stipulating that the contractor shall comply with all relevant Codes of Practice and British Standards, etc. Whilst this may be superfluous, because the standard forms require the contractor to exercise at least reasonable skill and care in the design, it may be considered, by some employers, prudent to include such a general statement, and in addition, to draw to the contractors' attention any special standards which may be applicable to the building, either general, e.g. nursing home registration requirements, or specific, e.g. conditions laid down in a lease agreement.

Circumstances Appropriate to Design–Build (Single-Stage Tender)

In practice, the Design–Build (single-stage tender) variety of Design–Build procurement, is probably the most commonly used option, because most employers seem to see it as giving them the dual benefit of controlling the design, so far, yet leaving the contractor scope to inject 'buildability' and new ideas into the design development phase.

As this is the case, employers wishing to use Design–Build will opt for this variety by default; that is, they use it unless, and only unless, the prevailing circumstances of the particular project dictate that one of the other varieties is more appropriate.

DESIGN–BUILD (TWO-STAGE TENDER)

Two-stage tendering is used in Traditional Contracting as well as in Design–Build, and it is important to contrast its use in these two forms of contracting.

Traditional Contracting (Two-stage Tender)

In Traditional Contracting, the two-stage tender enquiry would normally consist of a bill of approximate quantities together with some preliminary drawings. The competing contractors bid on the basis of rates against the quantities, and lump-sum prices for preliminaries and profit. The professional quantity surveyor will analyse the different combinations of rates and lump-sums and make recommendations as to which contractor's tender will lead to the most economic final cost; and the contract is agreed on that basis.

Meanwhile, the architect and the engineers will be developing the design so that construction work can proceed and, as the work is carried out, it is remeasured for payment to the contractor; that is, the tender rates against the new quantities, plus preliminaries and profit, as tendered. There may be variations to the rates and preliminaries if the design, as it is developed, causes there to be changes in the character of the work compared with that described in the approximate bill of quantities, and that which the contractor could have reasonably foreseen from the contract documents.

Obviously, as the design nears completion, which would normally be well before practical completion, the quantity surveyor could advise the employer as to the likely final cost. However, until the final account is agreed with the contractor, this is not a final figure.

It can be seen therefore, that with a two-stage tender, on a Traditional Contract basis, the employers have uncertainty of final cost at the time that they enter into contract and, to that extent, it is not a good system for them. Therefore employers only use it reluctantly, when they wish to use Traditional Contracting but time does not permit the normal pre-contract procedures to be adopted. In contrast we shall see that, given certain circumstances, a Design–Build two-stage tender approach could be of positive benefit to employers, and therefore they may choose it for its benefits, and not with reluctance.

Design–Build, Two-stage Tendering

Most Design–Build projects, including even Develop and Construct, have a negotiation period between the tender and the award of the contract, and to that extent they all have two pre-contract stages. However, with Design–Build (two-stage tender), employers deliberately frame their enquiries such that a second stage of design and pricing is inevitable before the contract can be agreed.

Contents of a Typical Design–Build (Two-Stage Tender) Enquiry

The enquiry information upon which employers invite tenders may, typically, consist of:

DRAWINGS
Architectural	Existing site plan
	Line drawing showing plan of accommodation
	(or neither of these)
Structural	Typically, none
Building services	Typically, none

SPECIFICATIONS
These would usually give a description of the accommodation which the employer requires.
   Particular specifications would only be included if the employer has definite ideas or requirements with respect to any particular elements of the project.

As employers have little to produce for these enquiries, they rarely engage a design team, and hence this is the first in the contractor-led-design category that we are considering. Employers may, however, engage consultants to assist in the evaluation process, and to check the contractor's design drawings as they are developed throughout the course of the contract. The employer's representation is considered more fully in Chapter 7, 'Employer's Representation'.

Circumstances Appropriate to Design–Build (Two-Stage Tender)

Generally, employers adopt this two-stage tender approach when they either cannot, or do not wish to, provide much information in the enquiry. There could be a number of reasons for this, for example:

(i) When an employer has insufficient time to prepare the enquiry documentation.

(ii) When an employer wants, what is, in effect, a costed design competition, which may arise in a number of circumstances, e.g.:

   (a) for a prestigious building suitable for a competition to find a design of distinction, style and quality, yet in circumstances where the employer wishes to have his costs, once agreed, protected against subsequent escalation by the provisions of a Design–Build agreement,

   (b) when a property developer who owns a site decides to seek competitive proposals to find the most cost effective solution,

   (c) when an industrialist has no particular preference for the design of a building to house his plant or equipment.

(iii) When there is a real doubt as to whether the project will proceed, and the employer wants to make preparations in the event that it does go ahead; but, at the same time he does not want to get involved with the expense of preparing a detailed enquiry, and likewise he is responsible enough to inform the contractors of the situation so that they will also not incur too much abortive expenditure in preparing a fully detailed tender.

(iv) When the project is particularly complex and the employer wants the design to be undertaken by the contractor's team, but first wants at least some competition introduced into the selection process, and a framework within which to control the second stage pre-contract negotiations.

Benefit of Design–Build (Two-Stage Tender)

As we can now see, Design–Build (two-stage tender) can be used by employers to obtain, without cost or obligation, a number of alternative design proposals, each with a relatively firm price and completion date attached. It is, in effect, a firmly costed design competition. This could be of obvious benefit to employers, and it contrasts with the disadvantages to employers of two-stage tendering in a Traditional Contract.

NEGOTIATED DESIGN–BUILD

Any contract may be negotiated, be it employer-led-design, or contractor-led-design, but here we are considering those negotiated contracts where the employer selects a contractor at the outset by means of a limited fee competition, or by means other than competition, to undertake the complete design and construction of the project. Throughout, from here on, we shall refer to this method of procurement as Negotiated Design–Build.

At first, it may seem strange for an employer to select a contractor without competition, but in practice many projects are undertaken in this way.

Enquiry Documents, Negotiated Design–Build

Normally, there are no enquiry documents, as such, but usually there are requirements of some sort given to the contractor by the employer and these may be adapted for use in a JCT 1981 contract as the Employer's Requirements.

Otherwise, the employer will not undertake any design work, and therefore he will not need any consultants except any that he may choose to employ to vet the contractor's proposals as they develop, and most importantly to assist him in negotiating the contract sum, bearing in mind that there may not be any comparable prices to help the employer establish whether he is getting good value for money.

Circumstances Appropriate to Negotiated Design–Build

There can be no doubt that the very act of seeking competitive tenders in any form will cost the employer both time and money, and therefore he should include this consideration in his decision making processes in addition to any other relevant factors or circumstances.

Examples of circumstances which may make it advantageous for employers to forsake competition, and to select a single contractor with whom to negotiate the contract include:

(i)   A follow-on contract, where an employer wants a building similar to one which the contractor has already designed and built satisfactorily.

(ii) Where a contractor has his own standard design for a building which may be regarded, in effect, as a 'product' which the employer may consider and 'buy'.

(iii) When a contractor has unique qualifications to undertake a particular project.

(iv) When the contractor has an interest in the building, financial, or otherwise.

(v)   When time is of the essence, so much so that an employer, whilst preferring Design–Build, has no choice other than to appoint a contractor of his choice at the outset to undertake the design and construction from start to finish.

Cost Limitation in Negotiated Design–Build

Before agreeing with a contractor to proceed to negotiate a contract, it would be normal and prudent for the two parties to have an understanding between them as to how the contract sum will be calculated and justified, and what safeguards the employer would have against exploitation. Equally the contractor should seek to safeguard himself from abortive costs should the employer abandon the project, or change contractors, part way through the design and negotiation stage.

Usually the parties will rely on measures which may include some of the following:

(i) Engagement of a professional quantity surveyor, by the employer, to vet the contractor's price build-up.

(ii) Agreement by the contractor to provide the employer with a detailed priced bill of quantities for checking. It may be a 'builder's bill of quantities' as opposed to one complying with the latest SMM (Standard Method of Measurement).

(iii) Agreement that all specialist sub-contracts, where practical, shall be the subject of competitive quotations, including any firms which the employer may select. The employer would have the opportunity to inspect the quotations, and they would be the basis of the price for those elements of the contract sum.

(iv) Agreement that the contractor will declare the detailed build-up of preliminary items.

(v) Agreement between the parties upon the formula for calculating the fees and profit levels. This would often be expressed as an agreed percentage of the contract sum, or as a lump sum calculated at a percentage of the original budget cost.

(vi) The appointment of the contractor at the outset by way of a 'letter of intent' which instructs the contractor to proceed, and includes the conditions to apply in respect of fees payable should the project be halted, or if for any other reason the contractor's appointment is terminated, through his fault, or otherwise.

It is not uncommon at a very early stage, perhaps even before the employer makes any firm commitment to pay fees, to agree a target cost, or a guaranteed maximum price, if there is sufficient information available upon which to base a realistic estimate.

DESIGN AND MANAGE

Is it Design–Build, or Management Contracting?

It is sometimes held that Design and Manage is not a form of Design–Build, but that it is a form of Management Contracting.

The question may not seem important, but under certain circumstances, it can have a significant effect on both contractors and the employers. Most medium, and large, contracting companies have separate divisions, and therefore different personnel for the two styles of contracting. Within the contractor's organisation the two divisions may well do battle over which should pursue and undertake a Design and Manage project, bearing in mind that such opportunities are generally attractive to contractors.

From an employer's point of view, difficulties may also arise. The employer who engages a contractor for a Design and Manage project will normally select a contractor with whom he has already enjoyed a successful working relationship. If that previous experience had been with one or other of the Design–Build or Management Contracting divisions, he may well get the division, with whom he has no experience, for the project simply because of the way in which the particular contractor has interpreted the nature of the proposed contract.

Two Forms of Design and Manage

In fact, there are two forms of Design and Manage, one which is Design–Build, and the other a form of Management Contract. The distinction lies in the nature of the contract into which the parties enter.

If at the culmination of the negotiations, the parties enter into a Design–Build form of contract, e.g. the JCT 1981 Form, then it would be considered as a Design–Build contract.

If the form of contract used is a Management Contract, e.g. the JCT 1987 Form, then it would be classed as a Management Contract.

For the reasons explained above, it is as well for employers to make their intentions clear at the very outset, before even approaching contractors.

We shall see in the next paragraphs how the Design and Manage negotiations may culminate in a Design–Build form of contract.

What is Design and Manage?

Design and Manage is similar, in a way, to Negotiated Design–Build in that the employer selects a contractor at an early date and employs him to undertake the majority, if not all, of the design of the project. However the distinction lies in:

  (i) The way that the contractor may be selected
  (ii) Method of procuring the construction work
  (iii) The way that the contract sum is calculated.

Contractor Selection for Design and Manage Contracts

An employer may select the contractor in the same way as he does in a negotiated contract. However, in Design and Manage he has the opportunity to inject an element of competition into the selection process. He can, at least, obtain quotations for the design and management fees. If the project is sufficiently definable or straightforward, then contractors could also be invited to quote for preliminaries, and (less likely) for the provision of common site services. An imaginative employer or his representative will have no difficulty in making the most of the opportunity for at least this level of competition, given the circumstances particular to the job in hand.

Method of Procuring the Construction Work in Design and Manage

Whether the project is defined as Design–Build or not, it is in this area that Design and Manage is similar to a Management Contract. All the construction work is undertaken by 'works contractors'; the main contractor normally provides site management, coordination and common site services, but nothing else.

Enquiries for works contractors' packages are prepared by the contractor, often in conjunction with a professional quantity surveyor appointed by the employer. The works contractors are selected on the basis of their responses to the enquiries and they enter into contract with the main contractor; in this context, the description main contractor is interchangeable with management contractor.

The Way that the Design and Manage Contract Sum is Calculated

There are two alternative ways in which the contract sum may be calculated on the Design–Build type of Design and Manage contracts. They are as follows:

(i) By use of PC (prime cost) sums;

A simple amendment to the JCT 1981 form of contract, could make it possible to list each of the proposed works packages, and against each a PC sum can be estimated.

The contract sum is then the sum total of the PC sums together with lump sums for the design and management fees, and for site preliminaries and the provision of common site services. As the works packages are established, the contract sum is adjusted for the difference between the PC sums and the quotations received.

In all other respects the terms of the contract would be in accordance with the standard JCT 1981 Form, or any other Design–Build form adapted suitably for the purpose.

(ii) By pre-contract negotiation;

The process would be the same as that described in alternative (i) for obtaining quotations and undertaking the work, but the works package subcontractors would all be established before the contract is signed, and hence firm prices rather than PC sums would be included in the original contract sum.

In this way the contract sum itself would be firm in the same sense as a normal Design–Build contract, and the JCT 1981 Form, for example, could be used without amendment.

The works contracts may include design responsibility, and whether or not they do, the main contractor remains responsible to the employer for the design of the works as a whole.

Circumstances Appropriate to Design and Manage

Employers usually adopt the Design and Manage alternative for one of the following reasons:

(i) In three of the examples of the circumstances where a Negotiated Design–Build would be appropriate, namely,

A follow-on contract similar to a previous one, or

where time permits little other choice, or

where the contractor has specific expertise

but the employer for his own reasons wants the whole of the construction work subjected to the disciplines of competition, and not just vetted by a professional quantity surveyor, as would be the case if it had been Negotiated Design–Build.

(ii) If, for any reason, the work cannot be quantified at the outset, and therefore a firm contract sum cannot be agreed before it is desirable, or necessary, to enter into the contract.

TURNKEY

The word 'Turnkey' has been in use for some considerable time, and it is fairly obvious that it is derived from the type of contracting where the contractor does everything, and all that the employer has to do is to 'turn the key', to use his building.

The employer, under a Turnkey contract, usually places the responsibility with the contractor for the complete design, construction, fit-out and commissioning of the building and its plant from concept to completion in response to a performance specification supplied by the employer for the performance of the works as a whole.

This concept and the term turnkey is traditionally applied to major industrial projects and they are frequently overseas, often in the developing countries.

Typical examples of such facilities include:

(i) Major industrial plants, e.g.

Steelworks

Aluminium smelters

(ii) Complete manufacturing facilities, e.g.
     Car plants
     Food manufacturing facilities
(iii) Major civil/mechanical works, e.g.
     Power stations
     Airports
     Transport systems
     Harbours and docks

Most of these projects are a far cry from the day-to-day building work which we normally regard as Design–Build, but they do nevertheless involve contractors in the design and the construction of the works, and to that extent they are Design–Build contracts. However this book is involved with UK Design–Build contracting and therefore we will dwell no further on Turnkey projects.

II The Employer's View

6 Which Form of Procurement?

'The Forms of things unknown'
Shakespeare [*Midsummer's Night Dream*]

KEY TOPICS

- Non-Design–Build methods of procurement
- Common forms of contract
- Choices of Design–Build procurement methods
- Circumstances influencing the choice of method, relating to: design, cost, time and other factors
- Examples of projects and alternative options
- Choice of form of Design–Build contract
- Guidance given by the JCT 1981 Practice Notes and the BPF Manual

INTRODUCTION

Employers who regularly procure construction work have their own experiences to draw upon in selecting how they shall go about procuring the next project, but occasionally they decide to try an alternative method. There are also many employers who have no experience to draw upon. Both must rely upon the advice of others, and their own common sense.

The increase in the alternative methods of procurement, and the different forms of contract, is a recent phenomenon and it has become a topic in its own right, inspiring books, pamphlets, seminars, lectures, on this topic alone.

In this book, a single chapter only is devoted to the selection of the procurement method, and therefore the coverage of the subject is selective and concentrates less on Design–Build in comparison to others, but on which variety of Design–Build procurement is likely to most suitable for different sets of circumstances.

NON DESIGN–BUILD METHODS OF PROCUREMENT

The most common non Design–Build methods of procurement are:

(i) Traditional Contracting, with several sub-types
(ii) Management Contracting

(iii)Construction Management
(iv)Lump-Sum (Plan and Spec) Contracting

Traditional Contracting

Broadly speaking a Traditional Contract is one where the employer appoints an architect, and other consultants, to be responsible for the design and specification of his project; and a professional quantity surveyor to prepare a bill of quantities upon which tenders, the contract sum and the final account are based.

The contractor has no responsibility for the design nor the accuracy of the quantities; he is paid for carrying out the work which he is instructed to do, and he is paid for what he thus carries out at the rates contained in the bill of quantities, or other rates if, in the event, the character of the work is not as described in the bill of quantities.

In Traditional Contracting, the contractor may be chosen by any one of the following means:

(i) competitive tender
(ii) negotiation
(iii)two-stage tender

Traditional Contracting, Competitive Tender

This has been, and still is probably, the most common method of building procurement. It is a widely understood in the UK and so it warrants no further description here; except to add that it is sometimes described as a 'lump-sum' fixed price contract. This is not so in practice.

Apart from changes requested by the employer, the contract sum may vary for a number of reasons, including the following:

(i) If there are any omissions or inconsistencies in the bill of quantities.
(ii) If, in the design development, the designers need to incorporate work not measured, nor appropriately described in the bill of quantities.
(iii)If, in nominating a subcontractor or supplier, the provision in the bill of quantities is inadequate.

Negotiated Traditional Contracts

In this method, the employer selects a contractor, other than by competition, and the professional quantity surveyor negotiates the contract sum with the contractor, on the employer's behalf, using the bill of quantities which he, the quantity surveyor, has prepared.

Once the contract is agreed, the post-contract procedures are the same as other Traditional Contracting methods.

Traditional Contracting, Two-Stage Tendering

Here, the employer engages the same team as he would in other traditional methods of procurement. Usually because of a shortage of time, the professional quantity surveyor prepares an approximate bill of quantities, or schedule of items. The contractor(s) are invited to price this bill of quantities or schedule, together with preliminaries and profit, and the contract is agreed on this basis. As the design develops, the professional quantity surveyor re-measures the work and the contractor is paid accordingly.

In essence, and in practice, the principal difference between this and the previous two methods is the degree to which the final account sum may vary from the contract sum due to change arising during the design development stage of the project.

Management Contracting

As the term implies a Management Contract is one where the contractor is appointed to manage the project, but not to build it. The employer and the contractor enter into an agreement with a fixed amount payable to the contractor for the provision of management, common site services and for the contractor's profit.

The construction work is undertaken by a series of works contractors, selected jointly by the employer's consultants and the contractor, usually on a competitive basis.

A Management Contract enables the employer to appoint the contractor at an earlier date than would be possible in most other methods. The contractor is usually selected on the basis of proposals that he puts forward for the management of the contract, and the fee that he quotes.

The proportion of the total cost represented by the Management Contractor's fee is relatively small, and so, even when he is in place the cost is not firmed up to any great extent. As the works contracts are let, the cost becomes increasingly certain, but due to the nature of this method of procurement some of the packages are not let until a late stage in the project. Perhaps, this lack of certainty is one of the biggest drawbacks of the procurement method.

Construction Management

Construction Management is relatively new to the UK, but similar arrangements are practised in other countries. With this form of procurement, the construction work is carried out by 'works contractors' engaged directly by the employer himself, and hence the employer takes on the contractual position of the main contractor. Since most employers do not have the expertise to manage the works contractors, they employ a Construction Management firm, on a fee basis, to do this on their behalf.

Firms that offer this service include both contracting organisations, and consultants, including professional quantity surveying firms.

The features of this method are similar to Management Contracting, except that the employer naturally has greater risks and problems if things do not go according to plan; on the other hand, if things go well he may get his project for a better price because he cuts out the contractor's profit.

Plan and Spec, Lump-Sum

This form of procurement is similar to the Traditional methods in so far as the contractor is not responsible for any design, either pre-, or post-contract. It is the employer's design team who are responsible and they complete the design in its entirety before inviting tenders. But unlike Traditional Contracting, the contractor is responsible for his own bill of quantities.

Because of the extent to which the design has to be complete before tenders can be sought, the method is rarely used, and then only on relatively simple projects.

This is almost a truly fixed price method; with one exception to the rule that variations only come about as a result of a change in the employer's requirements; and that is, if the design and specification, when they come to be used for construction, have to be amended for any reason; for example, a change needed for compliance with statutory requirements or, if the design itself includes any 'un-buildable' elements, errors or deficiencies.

NON DESIGN–BUILD FORMS OF CONTRACT

The standard non Design–Build forms of contract commonly used are as follows:

Forms which may be used for Traditional Contracts
    JCT Standard Form of Building Contract
    JCT Standard Form, with Approximate Quantities
    JCT Intermediate Form of Building Contract
    JCT Agreement for Minor Works
    JCT Fixed Fee Form of Prime Cost Contract
    ACA Form of Building Agreement
    ACA/BPF Form of Building Agreement
    GC/Works/1
    FAS Building Contract
    ICE Conditions of Contract
    ICE Conditions of Contract for Minor Works
Forms which may be used for Management Contracts
    JCT Standard Form of Management Contract
Forms which may be used for Lump-Sum (Plan and Spec) Contracts
    JCT Agreement for Minor Works
    ACA Form of Building Agreement
    GC/Works/2

CHOICE OF DESIGN–BUILD PROCUREMENT METHOD

Introduction

There are many projects or types of buildings which can undoubtedly be designed and built efficiently, economically and to the requisite standard of quality by either the Traditional Contracting method, Management Contracting, by Design–Build, or by any other form; providing always, that those involved perform well.

Evidence of this is the fact that there are employers who have similar building requirements, yet some choose one method, and others choose alternatives. One of the most obvious examples of this is in major retail construction, where at the time of writing, one of the largest such companies consistently uses Design–Build, another consistently uses Traditional Contracting and a third company normally favours Management Contracting. This is surely proof enough that one method is not in itself intrinsically better than another.

There are few, if any, examples of a number of employers with similar building needs all opting for similar procurement methods.

Some argue that the principal factors, which may lead an employer to follow a particular route, are as follows:

(i)   Is there a limit to the cost?
(ii) Must the cost be known accurately at the outset?
(iii) What is the likelihood that changes will be required?
(iv) Must competition be demonstrated?
(v)   Has the building to be to a high standard or prestigious?
(vi) How much control does the employer himself wish to exercise?
(vii) Is there a requirement for specialist designers?
(viii)Is time an overriding consideration?
(ix) Are there any special time constraints on parts of the works?

The problem here is that there are as many arguments for, as against, any particular procurement method with regard to most of these questions.

To sum up, it would seem to be fairly apparent that it is not the type of building, nor the type of business of the employer, nor even the prevailing circumstances which inevitably lead an employer to one procurement method in preference to another. The remainder of this chapter discusses which of the Design–Build varieties is likely to be most suitable in particular circumstances.

WHAT HAS DESIGN–BUILD TO OFFER?

More employers are beginning to see the advantages that a Design–Build route can offer them, and the growth in the proportion of building work undertaken in this way has been marked in the last few years and, for the time being at least, there seems to be no slowing down in this growth.

Rather than attempt to list those features which may appeal to the employers who now use Design–Build, we shall concentrate on which variety of Design–Build is likely to be most suitable. In doing this we will find that provided that the appropriate variety is adopted, many of the criticisms of Design–Build will be answered.

Chapter 5, 'Varieties of Design–Build Procurement', introduced us to the idea that there are a number of ways to procure work on a Design–Build basis; here we shall see what governs the employer's choice, once he has decided in the first place that he wants to use Design–Build.

To refresh our memories, the main varieties are:

(i)   Develop and Construct
(ii) Design–build (single-stage tender)
(iii) Design–Build (two-stage tender)
(iv) Negotiated Design–Build
(v)   Design and Manage
(vi) Turnkey

In Chapter 5 we briefly saw some examples of circumstances which would be appropriate to each variety; here we shall endeavour to answer the question of which variety to use in a more comprehensive and logical way. Rather than deal with the varieties in turn, and give features and examples as we did in Chapter 5, we shall look at different circumstances, and the different objectives that employers may have, to see if at least one or more Design–Build variety could suit him.

Circumstances Influencing the Choice of Method

Again, we remind ourselves that we are dealing here with the situation where the employer has already decided to use Design–Build, or at least, if he can find a suitable way of using Design–Build, then he will do so.

Circumstances which may influence the choice of Design–Build variety can be categorised into those relating to:

Design,
Cost,
Time, and
other particular circumstances.

Tables 6.1, 6.2, 6.3 and 6.4 show many possible circumstances within these categories, and suggest the varieties which each particular circumstance may suggest.

By checking the tables, a prospective employer may be guided to the most suitable method for his proposed project.

Table 6.1   Varieties of Design–Build suitable for circumstances related to Design

Circumstances relating to DESIGN	Possible Variety of Design–Build
The employer already has a design team in place	Develop and Construct Design–Build (single-stage tender)
The employer wishes to control the design team closely during the conceptual stage	Develop and Construct Design–Build (single-stage tender) Negotiated Design–Build
The building design is highly specialised and could only be designed by a specialist consultant, or the employer himself	Develop and Construct Design–Build (single-stage tender) Negotiated Design–Build
The building is highly specialised, but the expertise lies within contractor's organisations	Design–Build (two-stage tender) Negotiated Design–Build Design and Manage Turnkey
The building is to be prestigious (Note: The choice of contractor, designer and employer's representation are most important in this case)	Develop and Construct Design–Build (single-stage tender) Design–Build (two-stage tender) Negotiated Design–Build Design and Manage Turnkey
The building is one of a series of similar buildings	Develop and Construct Design–Build (single-stage tender) Negotiated Design–Build Design and Manage Turnkey
The employer predicts that there will be a substantial number of changes (Note: The contract documents should contain a detailed schedule of rates and the employer's rep. should have costing expertise)	Develop and Construct Design–Build (single-stage tender) Design–Build (two-stage tender) Negotiated Design–Build Design and Manage Turnkey
The employer wishes to instigate a costed design competition	Design–Build (two-stage tender)

Table 6.2   Varieties of Design–Build suitable for circumstances related to Cost

Circumstances relating to COST	Design–Build Variety
The employer wants to have a firm price before commitment to proceed	Develop and Construct Design–Build (single-stage tender) Design–Build (two-stage tender) Negotiated Design–Build Turnkey
The employer has a limited budget, and wants to know what is the best design concept that he can get for the money which he has available	Design–Build (two-stage tender) Negotiated Design–Build Design and Manage
The employer must be able to demonstrate that he has obtained competitive prices for the contract as a whole	Develop and Construct Design–Build (single-stage tender) Design–Build (two-stage tender) Turnkey
The employer can satisfy rules for obtaining competitive prices providing each element is the subject of competition	Develop and Construct Design–Build (single-stage tender) Design and Manage Turnkey

Table 6.3 Varieties of Design–Build suitable for circumstances related to Time

Circumstances relating to TIME	Design–Build Variety
The employer wants work to start on site in the shortest possible time	Negotiated Design–Build Design and Manage
The employer wants an early start but must have some form of competition	Design and Manage
The employer wants the project complete as soon as practicable, but cannot forsake competition	Design–Build (two-stage tender) Design and Manage Turnkey

Table 6.4 Varieties of Design–Build suitable for other particular circumstances

Other particular circumstances	Design–Build Variety
The employer has a complex set of approvals to obtain before the design can be frozen	Develop and Construct
The employer wants to have a design competition	Design–Build (two-stage tender)
The employer wants to employ a particular contractor	Negotiated Design–Build Design and Manage Turnkey
The employer acquires a site for development for which the previous owner had procured a detailed design and a bill of quantities	Develop and Construct Negotiated Design–Build Design and Manage
(Note, the employer may require the contractor to take over the consultants under a novation agreement.	

EXAMPLES OF THE USE OF TABLES 6.1–6.4 TO ESTABLISH POSSIBLE PROCUREMENT METHODS

Here we consider a few examples to see the sort of alternatives which an employer may consider for a given set of circumstances.

Example 1

Employer: Health Authority   Project: A District General Hospital
The employer would complete his analysis, typically, as follows:

Circumstance	Possible Design–Build Variety (See Tables 6.1 to 6.4)
The employer wishes to control the design team closely during the concept stage	Develop and Construct Design–Build (single-stage tender) Negotiated Design–Build
The building design is highly specialised and could only be designed by a specialist consultant, or the employer himself	Develop and Construct Design–Build (single-stage tender) Negotiated Design–Build
The employer wants to have a firm price before commitment to proceed	Develop and Construct Design–Build (single-stage tender) Design–Build (two-stage tender) Negotiated Design–Build Turnkey
The employer must be able to demonstrate that he has obtained competitive prices for the contract as a whole	Develop and Construct Design–Build (single-stage tender) Design–Build (two-stage tender) Turnkey
The employer has a complex set of approvals to obtain before the design can be frozen	Develop and Construct

Results of Check

The variety common to every 'circumstance' is Develop and Construct, and it is reasonably self-evident that this could well be the most appropriate procurement method for a district general hospital.

Design–Build (single-stage tender) comes a close second, and an employer may well choose this variety if the question of approvals could be resolved satifactorily.

Example 2

Employer: Manufacturer   Project: A production facility
The employer would complete his analysis, typically, as follows:

Circumstance	Possible Design–Build Variety
The employer predicts that there will be a substantial number of changes (Note: The contract documents should contain a detailed schedule of rates and the employer's rep. should have costing expertise)	Develop and Construct Design–Build (single-stage tender) Design–Build (two-stage tender) Negotiated Design–Build Design and Manage Turnkey
The employer wants to have a firm price before commitment to proceed	Develop and Construct Design–Build (single-stage tender) Design–Build (two-stage tender) Negotiated Design–Build Turnkey
The employer wants the project complete as soon as practicable, but cannot forsake competition	Design–Build (two-stage tender) Design and Manage Turnkey

Results of Check

First, we note that because of the likelihood of changes, the employer should take particular care in selecting his own representation, and also the contract documents should contain a detailed schedule of rates, to provide a basis for evaluating changes.

The varieties common to all the notional circumstance are:

Design–Build (two-stage tender), and Turnkey

Unless the project is of a particular type, the Turnkey option may well not be practical, thus, we are left with Design–Build (two-stage tender).

Develop and Construct and Design–Build (single-stage tender) were rejected on the grounds of speed because the pre-contract period, required for the preparation of enquiry documents, could well take more time than the employer could allow.

Negotiated Design–Build and Design and Manage were rejected because the employer wanted a firm price before proceeding.

Example 3

Employer: Property Developer   Project: Speculative Commercial
                                                 Development
The employer would complete his analysis, typically, as follows:

Circumstance	Possible Design–Build Variety
The employer wants to have a firm price before commitment to proceed	Develop and Construct Design–Build (single-stage tender) Design–Build (two-stage tender) Negotiated Design–Build Turnkey
The employer has a limited budget, and wants to know what is the best that he can get for the money which he has available	Design–Build (two-stage tender) Negotiated Design–Build Design and Manage
The employer wants the project complete as soon as practicable, but cannot forsake competition	Design–Build (two-stage tender) Design and Manage Turnkey

Results of Check

The variety common to all the 'circumstances' is Design–Build (two-stage tender).

Negotiated Design–Build, and Design and Manage come close seconds, and the actual circumstances prevailing at the time may persuade the employer to adopt one of these varieties.

Example 4

Employer: Hotel Operator   Project: A new hotel
The employer would complete his analysis, typically, as follows:

Circumstance	Possible Design–Build Variety
The employer wishes to control the design team closely during the concept stage	Develop and Construct Design–Build (single-stage tender) Negotiated Design–Build
The building is one of a series of similar buildings	Develop and Construct Design–Build (single-stage tender) Negotiated Design–Build Design and Manage Turnkey
The employer wants to have a firm price before commitment to proceed	Develop and Construct Design–Build (single-stage tender) Design–Build (two-stage tender) Negotiated Design–Build Turnkey
The employer wants the project complete as soon as practicable, but cannot forsake competition	Design–Build (two-stage tender) Design and Manage Turnkey

Results of Check

There is no option common to all the 'circumstances'. It is necessary therefore to check back on the alternatives to see if any of the constraints can be relaxed.

On re-checking the circumstances, we could get different suggestions as follows:

Time was considered to be a constraint, and, for this reason Develop and Construct and Design–Build (single-stage tender) options were rejected, but if the design of the hotel conformed to a largely pre-designed standard, then either of these options could be utilised, in this instance, without a time penalty.

Alternatively, the employer could review his need for competition. As the project is one of a series he is likely to know the economic project price based on previous contracts. For this reason he may choose negotiation as the fastest option.

FORMS OF CONTRACT FOR DESIGN–BUILD

Having chosen the variety of Design–Build which he wishes to use on a particular project, an employer must decide upon the terms of the contract to use for the agreement between himself and the contractor.

Reference should be made to Chapter 4 'Forms of Contract', where the salient features of the two principle standard forms of Design–Build contract are discussed.

The JCT 1981 Form is suitable for any of the varieties of Design–Build which we have considered in this chapter.

The BPF/ACA Form is normally only used when the employer is following the procedures of BPF System, and this implies that an employer-led-design approach should be used.

Before choosing the form of contract to use, employers are advised to familiarise themselves with the terms of the contracts and published guidance notes, which are mentioned later in this chapter.

If necessary, prospective employers should take further advice from experts or read more specialised publications on the forms of contract: there is no shortage of written commentaries upon the standard forms which guide the readers, clause-by-clause, through the terms and conditions.

Decisions for the Employer to make in Preparing the Contract

The two common standard forms of contract contain a number of points upon which the employer must decide, and these are:

JCT 1981 Form

    (i) Completion of the appendices
          contractual dates
          defects liability periods
          insurance terms and values
          liquidated and ascertained damages for delay
          fluctuation rules, if applicable
          payment terms
   (ii) Choice of optional/alternative clauses, in particular,
          insurances
          fluctuations

BPF/ACA Form

    (i) Insertion of particular details
          choice contract bills/schedule of activities
          client's representative
          times for various notices

payment terms
maintenance period
liquidated and ascertained damages for delay
(ii) Choice of optional/alternative clauses, in particular,
sectioning of the works
adjustments re contract bills/schedule of activities
insurances
law to apply to the agreement

Guidance Notes

Before attempting to make the decisions which are listed in the foregoing, an employer, or his representative is well advised to read through the various guidance notes issued by the pulishers of the forms of contract, as follows:

JCT 1981 Practice Notes

CD/1A
The practice note gives guidance to the employer on the following topics,
general use of the form
Employer's Requirements
development control requirements
building and accommodation requirements
data for conditions of Contract
exclusion of nomination of subcontractors and suppliers
the Contractor's Proposals
the Contract Sum Analysis
Value Added Tax
insurance, contractor's design liability
CD/1B
This practice note includes,
commentary on the Form of Contract
notes on the Contract Sum Analysis
application of formula adjustment

BPF Manual

This manual gives guidance to clients on the following topics
description of the BPF System
Stage 1, concept
Stage 2, preparing the brief
Stage 3, design development
Stage 4, tender documents and tendering
Stage 5, construction

responsibilities and duties
    client's representative
    design leader
    architect
    structural engineer
    building services engineer
    supervisor
checklists
BPF System forms

SUMMARY

In this chapter we first overviewed the non Design–Build forms of procurement. We saw that whilst arguments range over the features and alleged benefits of the various forms, there seems to be sound evidence that most projects could be undertaken quite satisfactorily using any of the common methods.

Given that an employer opts for a Design–Build approach, we considered how he may make his decision as to which of the varieties of Design–Build, introduced in Chapter 5, he might utilise.

Having chosen a variety, we saw how the employer may go about selecting the terms of the contract, and the guidance that is available to him.

The next fundamental decision, for the employer to make is, who should represent him, both in the pre-contract and post-contract stages? This is the subject of the next chapter.

7 Employer's Representation

'Your representative owes you,
not his industry only,
but his judgement'

Edmund Burke [*Speech in Bristol*]

KEY TOPICS

* Representation for employer-led-design projects

  Work stages
  Appointments to consider
  Stage-by-stage representation

* Representation for contractor-led-design projects

  Work stages
  Appointments to consider
  Stage-by-stage representation

* Summary charts showing representation for simple and complex projects

INTRODUCTION

All employers must decide what representatives and advisors they should employ for any given project. The need will vary, depending upon the employer's own expertise, the variety of Design–Build which the employer decides to adopt, and furthermore it may vary through the successive stages of the project.

Casual readers may find that this chapter appears to be unnecessarily repetitive, but they will find on closer inspection that there are subtle differences between the roles of the representatives and advisors at different stages and for different varieties of Design–Build.

To simplify the issue we shall consider the representation under two headings, namely:

(i) Employer-led-design projects
   Projects where the employer wishes to commission the design to a relatively advanced stage prior to the appointment of a contractor, as in
   Develop and construct, and
   Design–Build (single-stage tender).

69

(ii) Contractor-led-design projects
Projects where the contractor is largely, or completely, responsible for the design from concept to completion, as in
Design–Build (two-stage tender)
Negotiated Design–Build
Design and Manage
Turnkey

Project Stages

As the representation required by an employer varies through different stages of the project, it is necessary, for discussion purposes, to identify convenient stages of a project.

The BPF System, the RIBA Architect's Appointment and CAPRICODE each have their own definition of stages. Table 7.1 compares these stages.

Table 7.1   Comparison of the different work stages, defined by the RIBA, BPF and CAPRICODE

BPF Stages	RIBA Work Stages	CAPRICODE
1 Concept	A Inception B Feasibility	1 Approval in principle
2 Preparing brief	C Outline proposals	2 Budget cost
3 Design development	D Scheme design E Detail design F Production information G Bills of quantities	3 Design
4 Tender documents and tender	H Tender action J planning	4 Tender & contract
5 Construction	K Operations on site L Completion	5 Construction 6 Commissioning 7 Evaluation

The BPF System was devised by a panel representing property developers, and although the System is applicable to any final method of contracting, it is intended that the process should culminate in a Design–Build contract, and for this reason BPF stages are convenient for considerating the employer's representation.

EMPLOYER'S REPRESENTATION FOR EMPLOYER-LED-DESIGN PROJECTS

Appointments to be Considered

An employer must decide upon the personnel or firms which he will appoint for some or all of the following roles or disciplines. At some or all stages, one or more of the roles may be assumed by one person or firm.

(i) Project director
     The direct employee who is in a position to make strategic decisions.

(ii) Employer's representative
      The person who is referred to on a day-to-day basis for decisions.

(iii) Project manager
       The person or firm appointed by the employer to manage the project on behalf of the employer.

(iv) Design team leader
      He has overall responsibility for the design and other activities undertaken by the employer's design team (but not the contractor's design team).

(v) Architect
     The person or firm appointed by the employer to undertake architectural work his behalf, as opposed to any architectural work which the contractor is to undertake.

(vi) Structural engineer
      The person or firm appointed by the employer to undertake structural engineering consultancy on his behalf, as opposed to any such work which the contractor is to undertake.

(vii) Building services engineer
       The person or firm appointed by the employer to undertake building services consultancy on his behalf, as opposed to any such work which the contractor is to undertake.

(viii) Other specialist consultants
       The employer may wish to engage other specialist consultants to act on his behalf at any stage of the project.

(ix) Cost consultant
      Whilst it is not necessary to have a bill of quantities for a Design–Build contract, there are nevertheless, other costing services which a professional quantity surveyor, for example, could provide to the employer.

(x) Supervisor
     The employer may decide to employ person(s) or firm(s) to supervise the construction works, and to assist in inspections at practical completion and at the end of the defects liability period.

EMPLOYER'S REPRESENTATION, STAGE-BY-STAGE, EMPLOYER-LED-DESIGN

The comments on the following pages cover the employer's representation for each of the five stages of the project as defined earlier in this chapter.

Figures 7.1 and 7.2 (on pages 81–2) summarise the comments graphically for the different cases.

Employer's Representation – Employer-led-design

Stage 1 – Concept

In this stage the employer gathers as much information as he can, at a minimum cost commensurate with the nature of the project. His objective is to test the initial feasibility of the project prior to entering into any major commitments.

Project director
   His nomination is essential for this stage and all subsequent stages.
Employer's representative
   Usually there is no need for such an appointment at this stage, as the project director or the project manager assumes the role.
Project manager
   The project director often project manages the project at this stage, if his time and the nature of the project permits.
   At the concept stage, the project manager will often be a direct employee. Alternatively the design team leader or the architect may assume the role. If the proposed project has features demanding special management, the employer may appoint a specialist project management firm.
Design team leader
   It would not normally be necessary to make a formal design team leader appointment at this early stage, because the design work involved is normally minimal.
Architect
   Surprisingly often, it is possible to test the feasibility of a proposed project by costing it, without the need to produce architectural drawings. Consultants and contractors can usually estimate a square foot (or metre) costing for most types of regular building with sufficient accuracy for the employer's purposes at this stage.
Engineers – all disciplines and other consultants
   Unless the proposed project's feasibility is dependent upon the solution to an engineering problem, it would be rare to appoint an engineer at this early stage.
Cost consultant
   A professional quantity surveyor could be one of the most valued advisors at this stage.

Stage 2 – Preparing the Brief

Having decided in principle to proceed, the employer will move to the next stage, in which he causes the brief to be prepared. The stage does not go so far as to take the design into any detail, nor even to seek detailed Planning Permission but, if there is any doubt as to whether it is likely to be obtained, application for Outline Planning Approval would made during this stage.

The employer will see this as the second of three stages and, at the end of the third, the building will be almost completely design and specified. Usually this fact will influence his choice of representation for this stage so that he can move smoothly, into the next stage, on the back of the outline work done here.

Project director
> He will continue in his position as before, having given his instructions for the project to proceed to this stage.

Employer's representative
> His appointment at this stage will depend on the nature of the project. If not appointed here, the role will be fulfilled by either the project director or the project manager.

Project manager
> The employer should be relatively firm in his plans for the project management by this stage. If the nomination had not already been made, it should be now. However, the task may be delegated to a direct employee even if the employer intends to engage a project management firm at a subsequent stage.

Design team leader
> The employer should nominate the design team leader at this stage. Typically he will be the architect for a normal building project, or an engineer for a project where engineering is the predominant design skill required. Rarely will the design team leader come from a firm which is not involved in the design, of at least part of the project.

Architect
> The architect, who is to prepare the outline design, should be appointed at this stage, bearing in mind that the same architect will continue with the design development during the next stage.

Structural engineer and other consultants
> Normally the structural engineer's and other consultants' appointments at this stage will be limited to comments only upon the design of the works in relation to their own disciplines.

Cost consultant
> At this stage the first cost plan will be required, and employers normally appoint a professional quantity surveyor for this purpose.

Stage 3 – Design Development

This stage involves the design development to a point where, for example, detailed Planning Permission could obtained, and this stage also includes relatively detailed specifications for all finishes and engineering works.

Project director
   He continues to be needed for strategic or policy decisions.
Employer's representative
   At this stage, the work has become more intense, so it is essential to have an employer's representative for day-to-day decisions.
Project manager
   On medium to large projects, this now becomes a full-time role, often with more than one person in the project management team. An 'in-house' project manager often doubles as the employer's representative.
Design team leader
   The design team leader should have been established in the previous stage and so he continues, as before.
Architect
   The architect continues from the previous stage. His role and responsi-bilities must clarified, bearing in mind that it is not a Traditional Contract, and there may be a project manager and design team leader taking on administrative and management duties which the Traditional architect would otherwise undertake.
Structural engineer
   Again the scope of the structural engineer's work must be carefully worked out by the employer's representative and/or project manager. In Design–Build it is in dealing with structural matters that most people recognise that the contractor can influence the design to the best effect, and so he must be given as much scope as possible. At this stage, the structural engineer will give general advice only.
Building services engineers
   During this stage the building services engineer will be advising the design team on general requirements of the systems, spaces required and how the services may influence the design work of others.
Other specialist consultants
   These will be engaged to suit the design requirements of the job.
Cost consultant
   Assuming the employer has appointed a professional quantity surveyor he will constantly monitor the cost plan, advising the employer and the design team of the effect of changes.

Stage 4 – Tender Documents, Tendering and Contract

In the BPF System, Stage 3 should be completed and have received approval before Stage 4 is commenced. An employer may wish to adhere to this principle, or he may prefer to run this stage concurrently with the latter part of Stage 3 in order to save time. Stage 4 involves the assembly of the tender documentation, the definition of the contr'ct terms, the tender process and evaluation and the appointment of the contractor.

Project director
   Required as before.
Employer's representative
   Required as before.
Project manager
   Required as before.
Design team leader
   His role is to ensure that the design team now work together to produce a coordinated set of unambiguous drawings which accord with the latest decisions of the employer and are capable of being interpreted by the Design–Build contractors for their tenders.
Architect
   The architect, who may be doubling as the design team leader, brings his drawings into line as required above and, since it is these drawings in particular which will illustrate the works, he must ensure that they are sufficiently comprehensive.
Structural engineer
   He will normally be required to write the structural performance specifications, and prepare the documentation relating to ground conditions.
Building services engineers and other specialist consultants
   They will lay down comprehensive performance specifications, usually including the types of systems required and often manufacturer's names. There will be more detail in Develop and Construct, and less in Design–Build (single-stage tender).
Cost consultant
   The professional quantity surveyor will normally put together the contractual sections of the tender documentation, and lay down the format of the cost break-down to be supplied by the tendering contractors.

Stage 5 – Construction

This stage commences when the Design–Build contractor is appointed, and therefore there is a significant shift in responsibility at this point from the pre-contract design team to the contractor's design team.
Project director
   He continues, as before, but due to the shift in responsibility the nominee sometimes changes at this stage. For example, some companies may have a development directorate, responsible up to Stage 4 (or 3), after which their construction directorate takes over.

Employer's representative

This is the person who either makes decisions, or is the focus for decisions on the employer's behalf, and so his role is vital. He is often specifically named in the contract documents; the Client's Representative in the BPF/ACA Form, and the Employer's Agent in the JCT 1981 Form.

Project manager

Unless the project is unduly complex, or the circumstances of the job are such that there is a need for coordination with other events, or processes, it is rare for the employer to need a 'project manager' in addition to the employer's representative, already mentioned.

Design team leader

The form of contract will dictate how changes are conceived. If the employer wishes, he may employ his design team to carry out the initial stages of design for changes (reflecting the spirit of the design of the job as a whole), in which case he will retain the design team leader. The employer may also decide to pay to have the contractor's production drawings checked, or sanctioned, as they are produced. These decisions will dictate whether a design team leader is needed.

Architect

See the notes above regarding the design team leader; the need for an architect at this stage will be determined by similar decisions.

Structural engineer

The employer may retain the structural engineer for checking the contractor's structural design and drawings as work proceeds, and to monitor tests. Some would say that this is a needless duplication, and provided that the contractor's design team is reputable (which it should always be) this fact, together with the safeguards provided by the Local Authority Building Control, is sufficient to satisfy the employer's needs.

Building services engineers

For projects with a substantial services content, it may be prudent for the employer to pay for a building services consultant to check the contractor's designs and details, and to inspect the works on site and to witness tests.

Other specialist consultants

It is indeed rare to have any requirement for other specialist consultants at this stage.

Cost consultant

The professional quantity surveyor, if appointed, would vet the interim valuations, estimates for changes, advise on contractual issues, and make recommendations with respect to the final account. Note here that the final account negotiations should be far less complex than they are in a Traditional Contract. Due to the nature of Design–Build they normally consist of the original contract sum plus/minus authorised changes, and perhaps there may be fluctuations and liquidated damages for delay to consider, but normally, little else.

Supervisors
  The degree of supervision which the employer decides he needs will depend on the nature of the work, and it is usually provided, if needed, by the consultants described above, either on a visiting or resident basis.

EMPLOYER REPRESENTATION FOR CONTRACTOR-LED DESIGN PROJECTS

Appointments or roles to consider

The employer will need fewer advisors in a contractor-led-design project than in an employer-led-design project. They include:

Project director
  The direct employee who is in a position to make strategic decisions.
Employer's representative
  The employee or person who may be referred to on a day-to-day basis for decisions. Note that we dispense with a project manager, as his functions are assumed partly by the employer's representative in respect of the employer's organisation, and by the contractor for the overall management of the design and construction of the project.
Architect
  The employer may appoint an architect for preliminary design, to check the contractor's design and to supervise site works.
Structural and building services engineers
  The employer may appoint engineers to check the contractor's proposals, detailed design and to supervise site works.
Cost consultant
  Few employers have the in-house expertise to analyse the 'value for money' of contractor's proposals, especially when there is no similar offer from another contractor to make comparisons. This is not to say that an employer should always seek like-for-like tenders. Indeed we have seen elsewhere that it can be of positive benefit to an employer to have disparate proposals and thereby get the benefit of competition in design as well as in the pricing of construction work; and we have also seen when it may be wise for an employer to negotiate with a single contractor. An employer may engage a professional quantity surveyor to protect his interests in such cases.
Supervision
  The nature of the project will dictate the need for independent supervision. If needed, the employer would usually appoint one of the consultants previously engaged to check the contractor's proposals.

EMPLOYER'S REPRESENTATION FOR CONTRACTOR-LED-DESIGN PROJECTS

The comments on the following pages cover the employer's representation for each of the five stages of the project as defined earlier in this chapter, and on page 82, Figures 7.3 and 7.4 summarise the comments graphically for the different cases.

Stage 1 – Concept

In this stage the employer gathers as much information as he can, at a minimum cost commensurate with the nature of the project. His objective is to test the initial feasibility of the project, prior to entering into any major commitments.

The advisors he needs are minimal and often he can complete the stage without appointing anyone on a formal basis. Many contractors are willing, on bona-fide projects, to provide sound advice without commitment other than a gentleman's understanding of the benefits which will be granted to the contractor should the project proceed. However, the employer should nevertheless consider the possible appointments, as follows:

Project director
 His nomination is essential for this stage and all subsequent stages.
Employer's representative
 Unless the project director handles the project at this stage, an employer's representative should be nominated. Rarely will this be his actual title. Often he may be called a project manager, or simply be an engineer or development surveyor or another member of the employer's staff nominated to carry out the necessary duties.
Architect
 Surprisingly often, it is possible to test the feasibility of a proposed project by costing it, without the need to produce architectural drawings. Consultants and contractors can usually estimate a square foot (or metre) costing for most types of regular building with sufficient accuracy for the employer's purposes at this stage.
Engineers – all disciplines and other consultants
 At this stage, with these varieties of Design–Build, it is rare for the employer to need the services of an engineer or other consultant.
Cost consultant
 If the employer is obtaining his advice directly from a contractor at this stage, there would be no need to employ a cost consultant. Otherwise, if he has no other source of such advice, he may seek the opinion of a professional quantity surveyor so that he can either proceed to the next stage on a sound basis, or abandon the project.

Stage 2 – Preparing the Brief

Having decided in principle to proceed, the employer will move to the next stage, in which he causes the brief to be prepared. This is the final stage before the contractor is appointed, and generally the employer keeps the work involved in preparing the brief to the minimum which will allow contractors to make constructive and relevant proposals.

Project director

   He will continue in his position, as before, having given instructions for the project to proceed to this stage.

Employer's representative

   If not already nominated, the employer's representative should be at this stage. It is essential that the employer has someone competent, in this position, dealing with the project on a day-to-day basis.

Architect

   Normally the employer will not need an architect at this stage. Often, outline plans can readily be prepared by the employer's staff; especially, for example, with industrial projects or owner-occupier bespoke offices.

   If Outline Planning Approval is to be obtained, this may be a good reason to use an architect, unless someone else, who is already in the employer's team, is capable of making the application.

Engineers and other consultants

   It is rare for the employer to engage directly any engineers or other consultants at this stage.

Cost consultant

   At this stage the first cost plan will be required, and employers often rely upon the contractor, or appoint a professional quantity surveyor for the purpose.

Stages 3 and 4 – Scheme Design, Tender and Contract

The contractor becomes involved at the beginning of this stage, either after the first part of a Design–Build (two-stage tender), or by appointment as the contractor for a Negotiated Design–Build or Design and Manage project. His appointment will normally be on the basis of a letter of instruction to proceed with the design work, and there will usually be an agreement as to how the contract sum will be calculated and possibly limited. There may also be various other conditions agreed between the parties to cover various eventualities that could occur. The stage ends when the employer and contractor enter into contract.

Project director

   He continues to be needed for strategic decisions.

Employer's representative

   He is the point of contact with the contractor and he will give decisions and monitor progress. He will normally be responsible to the project director for ensuring that the contract documents are properly drafted and completed, so that they reflect the requirements of the employer.

Architect and engineers

The employer would not normally have a directly appointed architect in his team at this stage for relatively straightforward projects, but where a project is either large or has complexities of some sort, it may well be in the employer's interests to engage a small design team on a limited basis to check the work of the contractor's designers. The professions that are represented in the employer's team will depend on the nature of the project.

Cost consultant

During the design development stage the employer's professional quantity surveyor, if appointed, has little to do. His major task comes at the end of this stage, in assisting in the negotiation and agreement of the contract sum and the contract terms.

Stage 5 – Construction

In contractor-led-design projects, the construction work often starts, on the strength of a letter of instruction, before the contract is signed, and to that extent this stage is often concurrent with the completion of the previous stage.

Project director

He continues, as before.

Employer's representative

The employer's representative's role is primarily to give instructions or approvals to the contractor. His duties may be widened, depending on what other representation the employer has in his team.

Architect or other designers

If the employer had no consultants in his team up to this stage, it would be unusual to engage them now. Even if he had them, he may not need them to continue during the construction stage, as there may be nothing that they can sensibly contribute to the project. On the other hand, on some projects it may be clear that they do have an important contribution to make in the checking of calculations, working drawings and details and in making site inspections and witnessing tests, etc.

Cost consultant

If a professional quantity surveyor had already been appointed by the employer for the previous stage, it does not follow that his services are necessary for the construction stage. The employer will know how confident he is in protecting his own commercial interests once the contract has been set up, and if he has the confidence, and he does not foresee the need for advice in respect of valuing changes, then he would not employ a cost consultant for the construction stage.

Supervision
>The nature of the project, and the strength of the contractor's own quality control procedures, will dictate whether the employer should retain his own supervisors. If he needs to provide independent supervision he will normally choose consultants, if any, who were previously involved at an earlier stage of the project.

REPRESENTATION AFTER COMPLETION OF CONSTRUCTION – DEFECTS

The essence of Design–Build is that there is only one single point of responsibility, and therefore, if there should be any latent defects, it should be a simple matter to get the contractor to return to put the defect right. This may be so in the majority of cases, but from time to time, employers may suffer defects to which there seems to be no practical remedy, or deficiencies which the contractor alleges are not his responsibility.

In such situations, the employer should seek the professional advice of a consultant who was in his team; failing that, the professional quantity surveyor could give advice, if the employer had used one. If the employer had no representation throughout, then he should identify an independent firm with appropriate experience to assist him in the resolution of his problem.

SUMMARY

Figures 7.1 to 7.4 show in graphic form a summary of suggested employer representation for employer-led-design and contractor-led-design projects, and contrasts simple and complex projects.

In the next chapter we consider the terms and conditions for the appoinment of the employer's architects and engineers.

Figure 7.1   Employer-led-Design, Simple Project

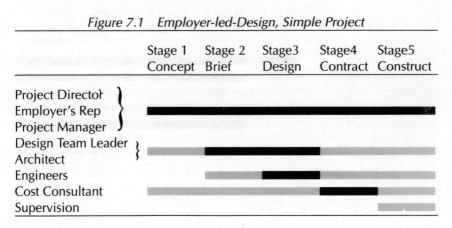

	Stage 1 Concept	Stage 2 Brief	Stage3 Design	Stage4 Contract	Stage5 Construct
Project Director Employer's Rep Project Manager					
Design Team Leader Architect					
Engineers					
Cost Consultant					
Supervision					

Figure 7.2   Employer-led-Design, Complex Project

	Stage 1 Concept	Stage 2 Brief	Stage3 Design	Stage4 Contract	Stage5 Construct

Project Director
Employer's Rep.
Project Manager   }
Design Team Leader
Architect
Engineers
Cost Consultant
Supervision

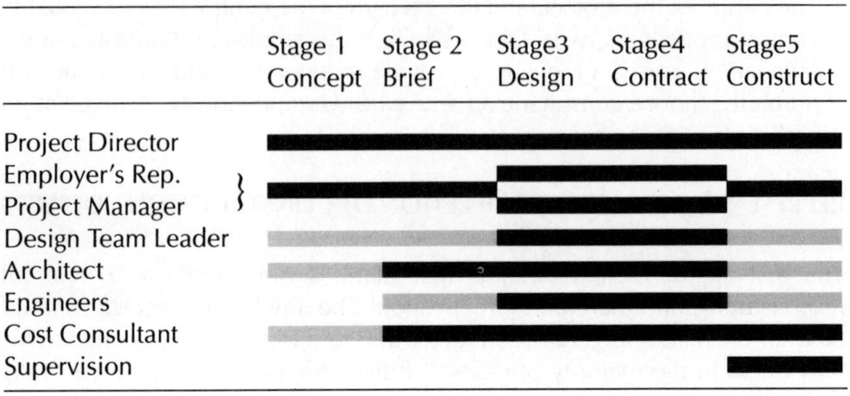

Figure 7.3   Contractor-led-Design, Simple Project

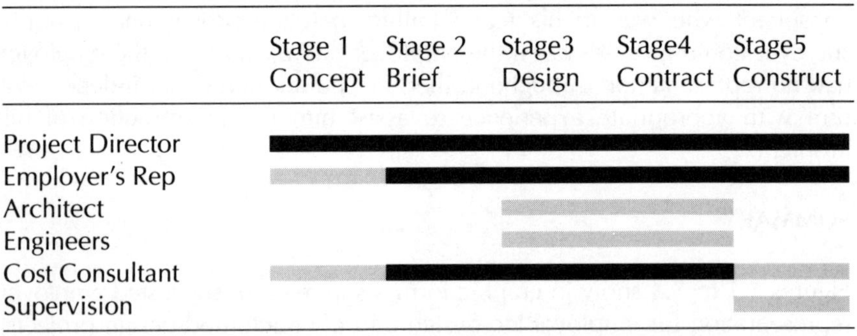

	Stage 1 Concept	Stage 2 Brief	Stage3 Design	Stage4 Contract	Stage5 Construct

Project Director
Employer's Rep
Architect
Engineers
Cost Consultant
Supervision

Figure 7.4   Contractor-led-Design, Complex Project

	Stage 1 Concept	Stage 2 Brief	Stage3 Design	Stage4 Contract	Stage5 Construct

Project Director
Employer's Rep
Architect
Engineers
Cost Consultant
Supervision

KEY   = Probable
       = Possible

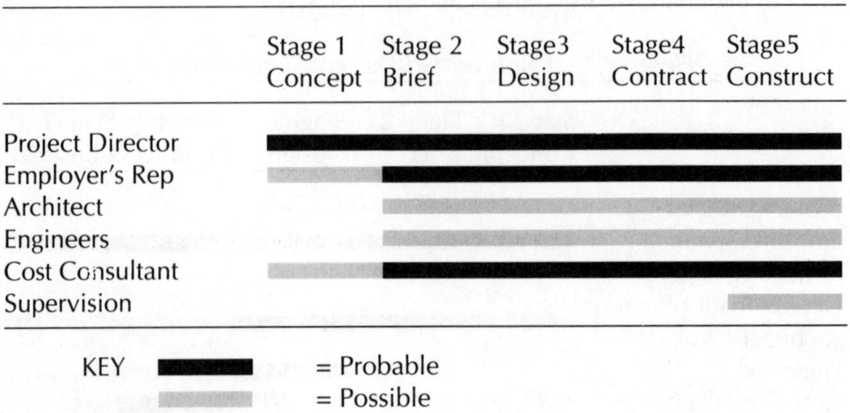

8 Employer's Consultants' Terms of Appointment

'No terms except unconditional and immediate surrender can be accepted'

General Grant [*To General Buckner*]

KEY TOPICS

- Architect's appointment:
  Conditions
  Services
  Fees
- Engineer's appointment:
  Conditions
  Services
  Fees
- Professional quantity surveyor's appointment

INTRODUCTION

There are standard forms which can conveniently be used for consultants' agreements in Traditional Contracting.

The RIBA's 'Architect's Appointment' is used by most employers for the architects' appointments, and the relevant ACE Conditions are used for most engineers' appointments.

These standard forms need some editing to make them suitable for the employer's use in the Design–Build situation and, of course, the services to be provided will vary depending upon the variety of Design–Build which the employer selects for the project. This chapter looks at how these standard forms may be used.

EMPLOYER'S ARCHITECT'S APPOINTMENT

Conditions of Appointment

The RIBA Architect's Appointment contain the following subjects within the conditions:

The duty of care
The architect's authority
The architect's obligations to keep the employer informed

The architect's relationship with other consultants
The architect's relationship with subcontractors and suppliers
Site inspections
Employer's instructions
Copyright
Assignment
Suspension and termination
Settlement of disputes
The governing law

Amendments to the RIBA Architect's Appointment

Apart from the exceptions noted below, the conditions are of a general nature which could apply to Design–Build as well as Traditional Contracting. The exceptions which may require editing are:

Architect's authority (Clause 3.4)
> Under this provision the architect is required to inform the employer if the building cost or the contract period are likely to vary materially, but in the Design–Build situation the architect will not often be in a position to fulfil this obligation; if this is the case the clause should be deleted or amended.

Relationship with other consultants (Clause 3.7)
> This clause gives the architect the 'authority' to coordinate the work of other consultants. This means in effect that the architect is the 'design team leader'. The clause should be edited, if necessary, to reflect the intentions of the employer, and he may wish to strengthen it by making it an obligation as opposed to the rather passive granting of the authority.

Subcontractors and suppliers (Clauses 3.8 and 3.9)
> Rarely, in a Design–Build situation, will the architect have any responsibilities with respect to subcontractors' or suppliers' design work, and therefore the clause may be deleted or, at least, edited.

Site inspections (Clauses 3.10 to 3.12)
> These clauses will only apply if the employer employs the architect in the construction stage, and wishes to include this responsibility within the services to be provided by the architect.

Architectural Services to be Provided to the Employer

We saw, in Chapter 7, the representatives that the employer may need for the various stages and types of project. This chapter considers the services which those representatives may be employed to undertake – particularly the architect and engineers.

Tables 8.1 to 8.5 illustrate the services, comparing the needs on employer-led-design with contractor-led-design projects.

Table 8.1   Employer's Architects' Services Stage 1 – Concept

Information to be provided or service to be undertaken by the architect	Employer-Led-Design	Contractor-Led-Design (provided an architect is engaged for this stage)
Take employer's brief	Always	Usually
Site appraisal	Always	Usually
Discuss with LA planners etc.	Often	Often
Advise on other consultants for this stage and/or others	Usually	Possibly
Produce outline plans	Usually	Usually
Give advice on cost	Occasionally	Occasionally

Table 8.2   Employer's Architects' Services, Stage 2 – Brief/Outline Design

Information to be provided or service to be undertaken by the architect	Employer-Led-Design	Contractor-Led-Design (provided an architect is engaged for this stage)
Further advice on the appointment of other consultants	Usually	Possibly
Outline Planning application	If required	If required
Draw up a brief for the building	Yes, in detail	Usually
Coordinate design by others	If required	Not normally required
Conceptual design drawings	Always	Usually
Further advice on cost	Occasionally	Occasionally
Advice on contractor selection	Often	Possibly
Advice on contractual matters	Often	Possibly

Table 8.3   Employer's Architects' Services, Stage 3 – Design Development

Information to be provided or service to be undertaken by the architect	Employer-Led-Design	Contractor-Led-Design (provided an architect is engaged for this stage)
Develop scheme design	Always	——
Coordinate design by others	Usually	——
Prepare performance specifications for specialist trades	Normally	Rarely
Seek detailed Planning Permission	Normally	——
Assist with land lease agreements and other legalities	Often	Occasionally

Table 8.4   Employer's Architects' Services, Stage 4 –
Contract/Tender Docs, Tendering

Information to be provided or service to be undertaken by the architect	Employer-Led-Design	Contractor-Led-Design (provided an architect is engaged for this stage)
Prepare enquiry drawings (see Chapter 5 for check list)	Always	_____
Coordinate design by others	Usually	_____
Assemble coordinated drawings and performance specifications	Usually	_____
Advise on contract matters	Usually	Often
Examine and advise on contractor's proposals	Normally	Normally

Table 8.5   Employer's Architects' Services, Stage 5, Construction

Information to be provided or service to be undertaken by the architect	Employer-Led-Design (provided an architect is engaged for this stage)	Contractor-Led-Design (provided an architect is engaged for this stage)
Check contractor's drawings etc.	Usually	Usually
Visiting site supervision	Usually	Usually
Outline design for proposed changes	Often	Rarely
Inspections on completion	Usually	Usually
Certify applications	Occasionally	Occasionally

Fees for Architect's Services to the Employer

Competitive Fee Quotations for Architectural Services

The surest way for an employer to obtain the keenest fees is to obtain competitive quotations from a number of architects; this applies to any type of project or level of service.

Before the employer can seek competitive quotations, he must define the services which he requires from the architect, and the terms under which they shall be provided. The employer may use Tables 8.1 to 8.5, read in conjunction with Chapter 7, for this purpose, and if he has already the benefit of advice from others, a professional quantity surveyor for example, he could assist the employer in defining the terms and services.

The enquiry to the architects must be clear on what is to be included within the fees, and what would rank for additional payment, and how such additional charges would be calculated.

Negotiated Fees for Architect's Services

When it is not practical to obtain competitive quotations from architects, or when the employer wishes to engage a specific architect, the fees must be negotiated.

This could be done by referring to the RIBA scales (which, incidentally, are not mandatory) and agreeing a reduction to take into account the reduced scope of the architect's work.

On an employer-led-design project, one might normally expect the fees to be between 50 per cent and 70 per cent of the scale fee. The precise amount will depend upon the nature of the project, and it will be, no doubt, the subject of bargaining.

On contractor-led-design projects the scope of the architect's services are reduced so much that the scales fees will be of no use in determining even the starting point for the fee negotiations. With this category of Design–Build, the architect may be reimbursed on an hourly rate basis, or he may agree with the employer a lump-sum based on an hourly rate calculation – provided the extent and nature of the services can be adequately described beforehand.

EMPLOYER'S ENGINEERS' APPOINTMENTS (STRUCTURAL AND BUILDING SERVICES)

The structural and building services engineers are considered together in this chapter, because, even though their respective works are different, the type of service they provide to the employer is largely similar on a Design–Build project.

Conditions of the Appointment

The ACE Conditions of Engagement contain the following subjects within the conditions:

Under the heading of General Conditions
Definitions
Duration of Engagement
Ownership of Documents and Copyright
Settlement of Disputes

Under the heading of Obligations of the Engineer
Duty of Care

Apart from the exception noted in the following, these conditions, being of a general nature apply to Design–Build as well as to Traditional Contracting.

Definitions:
The definitions assume that the contract is Traditional, with an architect appointed by the employer. The definitions should be

checked to ensure that there are no anomalies; for example, the 'Architect' is defined as the architect appointed by the 'Client' (and not the contractor's architect).

Services to be Provided by the Engineers to the Employer

It is rare, even in a Develop and Construct project, for a structural or building services engineer to undertake any detailed design work on behalf of the employer.

As we saw in Chapter 7 their work is usually confined to the following:

Employer-led-design projects
(i)   Advice at feasibility stage, if there appear to be crucial engineering problems
(ii) Advice regarding soils and topographic surveys, and surveys of existing structures, services, etc.
(iii) Investigations into availability of statutory service supplies
(iv) Performance specifications for the structure and building services works
(v)   Preliminary advice to the architect to enable him to progress the scheme design
(vi) Checking contractor's proposals and working drawings
(vii) Site inspections.

Contractor-led-design projects
(i) Checking contractor's proposals and working drawings
(ii) Site inspections.

Fees for Employer's Engineers' Services

The ACE Conditions contain detailed scales for recommended charges for engineers' services, but they give little guidance for fees for the services to an employer as envisaged by this and the previous chapter, and so the parties are left to utilise one of the following methods:

(i)   The employer obtains competitive quotations
(ii) Hourly rates are agreed and the charges are time-based
(iii) If the scope of the services can be defined with sufficient accuracy, the engineer may quote lump-sum fees, and provide the employer with a detailed break-down in justification of the amount.

EMPLOYER'S PROFESSIONAL QUANTITY SURVEYOR

Conditions of the Appointment

The RICS has issued a Standard Form of Agreement for the Appointment of the Quantity Surveyor (October 1983 Edition). It is a relatively simple form

which may be readily used by the employer and the quantity surveyor.
  The agreed services are to be described in the appendix to the Form.

Employer's Professional Quantity Surveyor's Fees

The fees scales suggested by the RICS are not appropriate for the Design-Build situation, and therefore they must be agreed on some other basis. As with the employer's engineers, one of the following methods may be used:

(i)   The employer obtains competitive quotations
(ii) Hourly rates are agreed and the charges are time-based
(iii) If the scope of the services can be defined with sufficient accuracy, the quantity surveyor may quote lump-sum fees, and provide the employer with a detailed break-down in justification of the amount.

9 Contractor Selection

'A man cannot be too careful
in the choice of his enemies'

Wilde [*Portrait of Dorian Gray*]

KEY TOPICS

- Importance of choosing the right contractor
- Identifying interested contractors
- First stage elimination process and questionnaire
- Short-listing contractors for interview
- Interviewing contractors
- Compiling the tender list
- Selection of a contractor for a negotiated project

INTRODUCTION

Often the 'expert' employer will know, from his own experience, which contractors to put on the tender list, or to negotiate with. Unless this is so, the employer should go through a selection process to make sure that the contractor who is finally chosen has the right credentials for the project.

There are normally up to four stages leading to the final selection of the Design–Build contractor, and these are:

(i) The long list of interested contractors
(ii) Short list of contractors to interview
(iii) The contractors who will tender
(iv) The final choice of the contractor, by competition or negotiation.

In this chapter we shall consider each of these stages, in turn; but first we shall consider, in more detail, the reason why making the correct decisions are so important. Throughout, where a task is ascribed to the 'employer', this may be read as the 'employer's representative', if appropriate.

THE IMPORTANCE OF CHOOSING THE RIGHT CONTRACTOR

In Traditional Contracting, the employer has separate agreements with the designers and the contractor, and it is often argued that if one performs poorly, the employer at least has the satisfaction of knowing that the other, if performing well, will 'pull the other through', and prevent what may otherwise be a 'disaster'.

There are others who would argue the reverse and say that if either party under-performs, then because of the adversarial nature of the Traditional Contract, the project will turn sour, and so there is twice the risk of problems when the designer's and contractor's agreements with the employer are separate.

Whichever point of view is true, is not important; what is important is that the employer who enters a Design–Build contract 'puts all his eggs in one basket' by entrusting the whole of the post-contract design and the construction process to just one organisation. He may get some comfort from various advisors, but the fact remains that if the Design-Build contractor fails to perform satisfactorily, there will be problems and difficulties which may have been avoided if a better or more suitable contractor had been chosen.

Another factor which makes the choice of the Design–Build contractor important, is the consequence of the contractor going into liquidation. This eventuality is difficult enough for the employer in Traditional Contracting, but in Design–Build the problems are doubled at least, especially if the contractor's in-house staff are responsible for the design.

IDENTIFYING INTERESTED CONTRACTORS

The first stage in the selection or elimination process, is to identify an adequate number of interested contractors. Typically, the first list of contractors will contain at least twelve names before an elimination process to reduce the number to say six contractors for interview.

The regular employer will, no doubt, have his own list of potential contractors and almost certainly both he and other potential employers will have received, on a fairly regular basis, approaches from Design-Build contractors interested in being given an opportunity to quote for work. However, successful contractors are less likely to advertise themselves through mail shots and the like, so the employer should be circumspect in relying entirely upon the contractors making approaches in this way.

The employer may see construction sites in his area, and make a note of the name of any contractors who appear to have tidy orderly sites, good quality of workmanship or design, and speed or efficiency of construction.

If the employer has advisors, he will, no doubt, receive advice from them.

If the employer is unable identify sufficient contractors in any of these ways, he may contact the Building Employer's Confederation (BEC) at 82, New Cavendish Street, London, W1M 8AD, for advice.

FIRST STAGE ELIMINATION PROCESS

The elimination process may be conducted by using a questionnaire.

The employer may write to the 'long list' of contractors with a brief questionnaire which gives the contractors basic information about the proposed project, and also asks the contractor questions about his company and its experience.

Figures 9.1a and 9.1b illustrate a typical such questionnaire.

QUESTIONNAIRE – PART 1

PART 1 BASIC INFORMATION ON THE PROJECT
1 Description of the Project, for example:
Four storey office building, with two storey wing, containing open plan offices, c80%, and cellular offices, including an air conditioned conference suite. Total floor area c 2,500 sq m.
Non traditional foundations, poor ground suspected, employer to provide soils survey, but interpretation will be at the contractor's risk.
Structure: to contractor's design.
External elevations: curtain walling and facing brick.
Internal finishes: normal standard generally, high quality reception and conference suite.
Services: fully serviced.

2 Site location
Give site address if possible, at least give general location. It would not normally be necessary for the contractor to visit the site at this stage.

3 Form of Contract
It is in the employer's interests to specify the form of contract at this stage, and to give details of any proposed amendments.
He should also include if possible certain options, for example, fixed price. Often the details for the appendices would not be established at this stage, but if the employer has any unusual requirement, for example, high liquidated damages, he should state them.

3 Number of tenderers
The employer should state the number of contractors who will be on the tender list, and he should then stick to this number.

4 Key dates
Dates should be given for:
   Notification to selected tenderers
   Issue of enquiry documents, and tender period
   Date for 'letter of intent' or order to proceed
   Date for site possession
   Completion date or contract period; or state if it is the contractor to decide.

5 Form of Enquiry
The employer should list the documents to be issued, or describe the nature of the enquiry documentation.

Figure 9.1a   Typical questionnaire to be sent to interested contractors

QUESTIONAIRE – PART 2

PART 2 INFORMATION TO BE PROVIDED BY THE CONTRACTOR

1 The Company:
Company name and address.
Address of local office, if applicable.
Executive (Director) responsible.
Contact: name, address, telephone and Fax numbers.

2 Experience:
Actual turnover, last year: Total
                                    Design–Build
Estimated turnover, current year: Total
                                    Design–Build
For each of 3 Design–Build projects completed within the last 12 months, give:
Employer's name
Brief project description
Value
Start and completion dates
Contact name for a reference

3 Design
Will design be undertaken in-house?
If part is to be by external consultants, provide their names and addresses.

4 Other information
The contractor should provide the following information or documents:
Latest annual company report and accounts
Pre-printed company brochures
External consultant's brochures if appropriate

5 The employer may think it appropriate to ask the contractor to sign a declaration such as the following, at this stage:

On behalf of .... I confirm that if we are selected to tender, we shall submit a bona-fide tender on the terms indicated in Part 1 of this questionnaire.

Signed....................................
Position..................................

Figure 9.1b   Typical questionnaire to be sent to interested contractors

What if Too Few Contractors Express Interest?

There has probably never been a time or place where no contractor is interested in quoting for work, provided he believes that he has a fair chance of securing the project at a fair price and under fair terms.

Nevertheless, the employer may identify a number of contractors and send out the questionnaires, only to find that none, or too few, respond. If this happens the employer or his advisors should ask themselves the following questions:

(i)   Is the proposed form of contract appropriate?

(ii)   Are there proposed amendments to the standard form of contract which contractors would regard as unduly onerous?

(iii) Are there going to be too many contractors tendering?

(iv)   Does the form of enquiry give the contractor too much to do, at risk?

(v)   Is there any reason for the contractors to think that the project is unlikely to go ahead, e.g. because of potential planning problems or perceived funding difficulties?

(vi)   Are the liquidated damages unreasonably high?

(vii) Are any stipulated time periods unrealistic?

(viii) Did the original list of contractors include a sufficient number of appropriate companies?

(ix)   Has the employer or his advisors gained a bad reputation among the contracting fraternity for any reason, justified or not?

(x)   Has the project been priced already by one or more contractors?

Having established the probable cause, the employer could make whatever changes are necessary, and send out questionnaires amended as appropriate, or he may have found that the variety of Design–Build he had chosen was inappropriate, in which case he should consider an alternative variety of Design–Build. He may even consider an alternative method of procurement altogether.

DRAWING UP THE SHORT-LIST OF CONTRACTORS TO INTERVIEW

From the brief questionnaires received from interested contractor, the employer should be able to draw up a short-list for interview. In making the short-list, the employer's objective is to reduce the list to no more than say the six companies, who should be, apparently, the most likely make the best tender proposals, and then, if selected, undertake the design and construction work economically, and well, in terms of both time and quality.

Making the choice from the number of questionnaires received may appear to be a confusing task, and if this is the case then the employer may decide to apply a systematic approach to his decision process.

The most common way to achieve this, is to draw up a list of appropriate factors and to score each contractor marks, out of say 10, for each factor, and then apply rules, such as:

eliminate any contractor scoring less than 2 for any factor
otherwise pick those with the highest aggregate score.

The employer, if he is of a mind to, may give certain factors a 'weighting' by scoring them out of more or less than 10, depending on the perceived relative importance of the factors.

A typical score sheet is shown in Table 9.1.

Table 9.1 Typical results from scoring subcontractors' responses

FACTOR	MAX	A	B	C	D	E	F	G	H	I	J
		CONTRACTORS' SCORES									
General impression	10	6	9	5	7	8	7	4	2	4	2
Size	6	2	4	2	8	6	3	4	1	3	3
D–B experience	20	18	12	15	5	1	16	12	8	7	10
Recent experience	10	5	6	5	1	4	8	4	3	4	5
Designer's	20	14	9	7	16	17	12	5	3	5	6
Office locality	10	3	7	4	6	6	5	9	1	3	8
TOTAL	76	48	47	38	43	42	51	38	18	26	34

The ratings in, descending order, are as follows, with * marking the chosen firms:

```
*  F  51
*  A  48
*  B  47
   D  43   – eliminated through lack of recent experience
   E  42   – eliminated through lack of Design–Build experience
*  C  38
*  G  38
*  J  34
_____ cut-off line
   I  26
   H  18
```

Unless the employer wished to reconsider the elimination of D say, who scored well except on recent experience, he would be left with

contractors F, A, B, C, G and J

to progress to the next stage, which is the interview stage.

If the employer wished to have a reserve contractor, he would either choose I, or he may look again at those eliminated.

CHOOSING THE CONTRACTORS TO TENDER, THE INTERVIEW STAGE

In all but straightforward cases, the employer, or his representatives should interview all the (six) contractors. These interviews, normally called pre-qualification interviews, are distinct from interviews where the contractor makes a presentation of his proposals to the employer prior to the award of the contract.

The employer may wish to send out a more exhaustive questionnaire, especially if his project is of a challenging nature and will demand the highest level of resources, skills and commitment from the contractor.

The interviews will be most fruitful, if the employer gives the contractor prior warning of the agenda for the pre-qualification interview, and normally it will be of the same format as the detailed questionnaire.

The employer should ask the contractors to ensure that the relevant disciplines or departments of their organisations are represented, including the designers, whether they are in-house, or independent.

Topics for the Prequalification Interview, and Detailed Questionnaire

Company Size

The size of the contractor could be important to the success of the project. Obviously, the size of the proposed project should not be disproportionately high in relation to the contractor's annual turnover, nor to the turnover of the contractor's local office, if relevant; on the other hand, the project should not be so small, in relation to the contractor's normal type of work, that the contractor would be unlikely to submit a competitive tender.

Experience in Design–Build Contracting

It is important that the contractor understands all the implications of Design–Build contracting, and has the resources and skills required of him to manage and undertake the work, within the constraints of this form of contracting.

We see in Chapter 11, 'Contractor's responsibilities/risks', that the contractor takes on many risks which could cause him loss and expense without the chance of recovery, and which do not exist in other forms of contracting.

Without tendering experience the contractor is likely to under-provide through errors and omissions, and whilst this may appear to be to the employer's advantage, in that he will receive lower quotations, it rarely works this way in practice. A contractor who is losing money will almost invariably seek ways to redress the situation, and whilst in theory he has few opportunities in Design–Build, that may not stop him trying. Furthermore, quite simply, a contractor who is losing money, rarely performs well.

During the design and construction stage, the inexperienced Design-Build contractor is unlikely to manage the design team efficiently. He may expect too much, or too little, of them. He may over-manage them, imposing too much upon the design, so diluting the level of genuine expertise being applied to the design. This would be to the detriment of the project. Or he may under-manage them, and lose proper control of the design and coordination and as a result suffer delays and unnecessary cost escalation.

Coordination between the specialist subcontractors and the design team is a difficult but essential operation, and here again the inexperienced Design–Build contractor will have great difficulties.

An employer should therefore be wary of employing a contractor without the relevant Design–Build experience. This may seem unfair to such contractors because they will suffer the 'chicken and egg' problem – without experience, they will not secure the work, but without the orders they will never gain the experience. The employer may ask himself whether he should he be the 'guinea-pig', and probably, he will say no.

Experience in Construction of Similar Buildings

In most contracting companies, the 'construction' and the 'design' teams are separate, and the construction teams build under whatever form of contract applies to the particular project in hand. Thus, they acquire an expertise in construction of many types of building, though not necessarily Design–Build in each case.

Whilst it is preferable to have a contractor who has previously designed-and-built similar works successfully, it is often difficult to find one who has also got the other prerequisites. This is particularly true in sectors where Design–Build has rarely been used hitherto, for example in hospital or prison building.

Thus, provided the other selection criteria are satisfied, it would usually be safe to select a contractor who has the requisite experience of constructing buildings nf similar nature, scope and size, though not necessarily under a Design–Build form of contract.

Design Experience

It is essential in the selection process for the employer to take into account the competence of the design team. He must, therefore, know which company, firm or department that the contractor intends to employ for the design of the project.

Their relevant experience must be examined, and weight given to the design disciplines which are most important to the job. For example, if the building is heavily serviced, then the building services designers will be crucial to the success of the project; if the proposed building is to be a 'show-piece' demanding the best quality design, specification, detailing and workmanship, then the employer should be wary of utilitarian contractor's designers, or external consultants whose experience is confined, say, to housing or basic industrial work.

It is probably not important for the designer to have experience in design of similar buildings to that proposed, actually within the framework of Design–Build, but it is of much more importance that the designer has experience first in work of a similar nature to that proposed, second, that he has general experience of working for a Design–Build contractor.

Location of the Project

It is just as important in Design–Build as in Traditional Contracting that the contractor knows the area well; perhaps even more so. After all, the Design–Build contractor has so much more to manage than his Traditional counterpart. To give him the additional problem of working in unknown territory could well give him difficulties that he would not be able to manage; that is, unless other factors do not assume a greater importance.

Available Resources

The employer should satisfy himself that the contractor will be able to call upon the requisite resources for both the design and the construction of the works at the time when they are needed. This may be difficult to determine without questioning representatives of each of the firms or departments involved on the subject of current work-load, resources available and orders anticipated.

At the interview, the employer will make the subjective judgements concerning personalities and empathy with the organisations. On the objective questions the employer may make use of a scoring system similar to that used at the earlier stage. The scoring should be done following each interview, otherwise the employer will become confused with what has been said by the different contractors.

By these means employers can make up a final tender list, and it may be prudent to write to the chosen contractors not only to inform them of their inclusion on the list and the dates for the tender period, but also to ask them to reaffirm their intention to submit a bona-fide tender.

SELECTION OF A CONTRACTOR FOR A NEGOTIATED CONTRACT

The processes described earlier in this chapter apply to the selection of contractors up to the competitive tender stage. In a negotiated contract there is no such stage, but unless the employer has already identified the contractor with whom he wishes to negotiate the contract, he will have to choose from those that are interested. The procedures described in this chapter could also be used for this purpose.

It is perhaps appropriate here to look at the reasons why an employer may forsake the benefits of competition in favour of a negotiated contract.

Reasons for Opting for a Negotiated Contract

In Chapter 6, we considered how an employer may select the variety of Design–Build which would be most suitable for the project which he proposes to build. The format of the chapter was such that the benefits of negotiation were not summarised together. The benefits can be identified, by implication, from the following list of circumstances which may lead an employer to choose to negotiate with one particular contractor.

Circumstances Favouring Negotiation

(i) When a contractor is offering a pre-designed or standard building; even though an employer may be able to obtain competitive prices for similar buildings from other contractors, the fact that the building is a contractor's standard building precludes the employer from adopting the normal competitive processes.

(ii) Speed: an employer, himself, will know the importance of the completion date for any particular project and, if its importance outweighs the need to test that he is obtaining the lowest price through competition, then he will negotiate with a single contractor.

(iii) Repetition: an employer will recognise whether it is in his interests to stay with a contractor with whom he is happy, who has already worked for him on previous project(s) with a successful outcome.

(iv) Convenience: an employer may choose to negotiate with a single contractor if circumstances, at the time, make it convenient for him so to do, provided he feels that he has adequate safeguards and control over the contract sum.

(v) Quid-pro-quo 1: a developer-employer is often in a situation when his bid for land or property is more likely to be successful if he has detailed development proposals and a guaranteed price for the design and construction of the project.

(vi) Quid-pro-quo 2: a property developer, owning development land, may have a situation where he needs a design and price for the building work to interest a prospective purchaser. The developer and contractor may agree that the contractor provides the proposals and price, and is awarded the contract on a negotiated basis, providing the proposals are accepted by the prospective purchaser.

(vii) Quid-pro-quo 3: occasionally a supplier of goods or services to the contracting industry may negotiate a Design–Build contract with a contractor in exchange for reciprocal business.

(viii) Funding support: a contractor often has more substance than a property developer, and as such has access to funding which is not available the developer. In such circumstances the two parties will often agree to negotiate the contract in exchange for the financial support or guarantees.

This list, long as it is, is not exhaustive, but it does serve to demonstrate how often that competition may not be the answer to the way in which the employer should select his contractor.

FINAL CONTRACTOR SELECTION

The final selection of the contractor who will undertake the works will depend on the evaluation of the tender proposals, and this is the subject of the next chapter.

10 Evaluation of Contractor's Tender Proposals

'A man who knows the price of everything
and the value of nothing'

Wilde   [*Definition of a cynic*]

KEY TOPICS

- Evaluation of contractor's proposals, employer-led-design projects

  Compliant tenders
  Non-compliant tenders

- Evaluation of contractor's proposals, contractor-led-design projects
- Systematic checking of proposal documents

  Contract terms and conditions
  Precedence of documents
  General impression
  Programme
  Technical content

- Follow-up questions to tendering contractors

FINAL CONTRACTOR SELECTION

By now, employers should have gone through a selection process designed to ensure that any of the competing tenderers would be suitable to design and build the proposed project, and hence many, if not all, of the questions of the contractor's suitability and resources will have been answered already. So, unless anything has happened during the intervening period which leads the employer to doubt the contractor's suitability, then the employer need not go over the same ground again.

Notwithstanding this, the contractor should be in a position, by this stage, to identify the key personnel whom he intends to employ on the project, should he be awarded the contract.

This then means that the final selection will be made on the merits of the contractor's tender, both in design, price and any other factor which is relevant to the particular project, subject to the employer being satisfied with the proposed contractor's personnel.

101

EVALUATION OF THE CONTRACTOR'S PROPOSALS (EMPLOYER-LED-DESIGN)

We have seen already that the objective in both Develop and Construct and Design–Build (single-stage tender) is to obtain tenders which are based upon a detailed set of enquiry documents, and provided the bids received are compliant with the terms and conditions of the enquiry, then the employer should, in theory, be able to accept the lowest tender.

The first aspect of each contractor's submission to check, is whether the tender complies with the instructions given to tenderers in the enquiry, as follows:

(i) Compliant Tenders

    (a) Is the tender an unequivocal offer to undertake the works in accordance with the terms of the enquiry documents? If the answer to this is yes for any of the tenders received, then they can be set to one side for the time being, whilst the others are checked.

(ii) Non-Compliant Tenders

    (a) These tenders should be examined for their degree of non-compliance, and the employer should perhaps look for the following points:

        (i) Is the non-compliance merely a technicality which can be resolved simply, with or without financial adjustment?

        (ii) Is the degree of non-compliance so great that it is equivalent to an 'alternative tender'?

        (iii) Does the tender contain a rider saying in effect that the offer submitted is for the works, as described within the tender documents, and in the event of there being any discrepancy between this and the enquiry, then the tender shall take precedence?

When these questions are answered, the employer will be able to decide whether any of the non-compliant tenders should be invalidated. Those that are not invalidated, go forward for evaluation.

The evaluation of the 'alternative tenders', if any, will normally be made following a process much the same as that described for the contractor-led-design tenders later in this chapter.

For the compliant, or near-compliant tenders the assessment of the tenders will normally be on a monetary basis, with consideration also being given to other factors which may have been included in the enquiry, for example:

(i) depending on the nature of the project, life-cycle costing, which can be simple, or complex, and can include replacement costs of plant and various maintenance costs. It may also take into account predicted energy costs

(ii) alternatives for particular elements

(iii) programme

The balance between the tenders should be relatively easy for the employer or his representatives to assess on employer-led-design projects, and so a list of the tenders may be drawn up in order of value for money.

The other factors for the employer to consider are:

(iv) comparative cash flow requirements, which could have a significant effect on the project cost

(v) the personnel to be employed by the contractor and his method statement for the construction of the work.

Most experienced employers will not commit themselves to their final choice, without first interviewing all the (three) tendering contractors. Usually this takes the form of presentations by the contractors, and often each contractor determines the agenda for his presentation, allowing the employer to raise questions. At this presentation the employer can assess the merits of the contractors' personnel (design and construction), and the methods that the contractors propose to use for the construction of the project.

EVALUATION OF THE CONTRACTOR'S PROPOSALS (CONTRACTOR-LED-DESIGN)

Contractor-led-design proposals fall into the category of either:

(i) competitive tenders as in
  (a) Design–Build (two-stage tender), and
  (b) Design–Build (single-stage tender), when an 'alternative' proposal has been submitted by a contractor and it is acceptable to the employer, in principle, or

(ii) non-competitive proposals, as in projects which are
  (a) Negotiated Design–Build, or
  (b) Design and Manage

The proposals in both of these categories should be systematically checked to ensure that the they meet the employer's needs and, in the case of the competitive tenders, to decide which offer suits the employer best, by whatever measure is appropriate for the particular project.

The following comments and check lists include the scope of the questions that an employer may raise.

On straightforward projects many of the points may be superfluous, and the employer will decide for himself what is important to him.

On more complex or demanding projects, the employer will normally ensure that if he cannot appreciate the significance of the answers to the questions himself, he will seek independent advice.

SYSTEMATIC CHECKING OF THE PROPOSAL DOCUMENTS

The contractors' submissions should be checked systematically, and the following list of questions is typical for such an analysis.

General Questions

(i) Contract terms and conditions

Has the contractor accepted the proposed terms and conditions in full? If not, what qualifications has he made, and are they acceptable?

(ii) Precedence of documents

Are the terms of the offer such that it takes precedence over the enquiry, or furthermore, is the offer limited to the scope and nature of the work contained within the proposals, irrespective of the requirements contained within the enquiry?
If this is the case how will the employer ensure that his requirements are ultimately met?

(iii) General impression

Do the proposals give confidence that the contractor has understood the requirements of the project in the fullest sense, or do they appear to leave questions and difficult points unanswered, such that problems are likely to be encountered later?
Has the contractor abided by any instructions given in the enquiry as to the manner of the presentation of the proposals and the information to be contained therein?

(iv) Programme

Is the contractor's programme well considered and apparently realistic for the lead-in, construction and commissioning periods?
Does the programme take into account all the requirements for sectional completion and early access for others employed directly by the employer?

Detailed Technical Check of Contractor's Tender Proposals

(i) General

Has the contractor included any statement that could imply that he has not included the costs of full compliance with all such requirements as:

the latest Codes of Practice, British Standards, etc.,
the recommendations of any other relevant trade organisation,
all statutory requirements,
requirements of the local Building Control Officer,
the Fire Officer's requirements,
the requirements of any registration body (e.g. for nursing homes and any other requirement particular to the project)?

(ii) Sub-structure

Who takes the risk on ground conditions, including,
load-bearing capacity of the soil,
depth of the load-bearing strata for foundations and ground bearing slabs,
presence and depth of ground water,
presence of underground obstructions and rock, etc.,
presence of contaminated soil or other deleterious materials,
presence of mine workings, pits, wells, soft-spots, sink-holes, etc.,
depth of topsoil,
suitability of excavated soil for back-filling and re-use?
Is the nature of the proposed sub-structure work described adequately for the purpose of evaluation (notwithstanding that this may be of no direct interest to the employer)?
What is the nature of the water-proofing of any underground structures?
Do such precautions appear adequate, or on the other hand are they over-precautious, as may happen if there were no strict cost constraints as could be the case in a negotiated contract; or, incidentally, in a Traditional Contract?

(iii) Superstructure

Is the nature of the structural work adequately described?
Is the finish of exposed parts of the structure satisfactorily specified?
Will the contractor avoid the use of potentially deleterious materials, such as woodwool slabs, asbestos, sea-dredged aggregates, etc. which may lead to problems later?

(iv) Roofing

Do the proposals provide a sufficiently durable roof covering?
Will the contractor offer any extended guarantees on the roof covering?
Are access-ways for maintenance of roof-top plant and for general purposes included in the proposals, and do they provide adequate protection to the roof covering?
Will there be any acoustic problems associated with a lightweight roof structure, e.g., close to a flight path of low flying aircraft?
Are the provisions for safety adequate?

(v) Stairs

Is the staircase construction specified, and is it suitable?
Are the floor, wall and soffit finishes specified for each staircase, and are they suitable?
Is the balustrading specified, and is it acceptable for all staircases, especially the main staircase?

(vi) External walls

If a PC sum has been included for facing work, is it realistic?
Are the aesthetic qualities of the external facades satisfactory?
Could there be acoustic problems?

(vii) Windows and curtain walling

Can the type of window or system, including ironmongery, be identified from the proposals, and is the specification satisfactory?
Is the glazing specified and will it be functionally and aesthetically suitable?
Has the contractor checked for the possible incidence of condensation?
Are there any features which will cause detailing problems and lead to later defects?
Are the windows acoustically satisfactory?
What provisions are included for cleaning windows and external facades?

(viii) External doors

Is the specification for all external doors clear, and will they be satisfactory in each case?
Are roller shutters, etc. to be manually, or electrically, operated?

(ix) Internal walls and partitions

Can the proposed form of construction of all internal walls and partitions be identified, and are they suitable in each case for flexibility robustness, durability, acoustics and general strength?
Are the walls suitable for any employer's fittings that are to be fixed at a later date?
Are there any changes in the form of construction at undesirable places, such that there is a risk of movements cracks appearing later?
Do walls go up to the soffit where necessary for acoustic separation between rooms?

(x) Internal doors

Is the specification clear as to the type of door and frame in each case, and are they suitable from the point of view of cleaning, durability, robustness, appearance and function?

(xi) Wall, floor and ceiling finishes

Has the contractor provided a comprehensive room finishes schedule? If not, how are the finishes specified and are the descriptions adequate?
Are the finishes suitable, in each case, for wear-and-tear, durability, corrosion appearance, hygiene, cleaning, maintenance, safety, static electricity, acoustics, etc?
Are such things as tiled surfaces adequately described?
If a PC sum has been included for any material is it realistic?

(xii) Fittings

Are any fittings to be supplied by the contractor, and if so, are they scheduled and do they meet the requirements?

(xiii) Sanitaryware and appliances

Is all sanitaryware adequately described, and is the employer satisfied with the proposed appearance and type in all case?

(xiv) Plumbing

To what extent will pipes be exposed?
Are any pipes to be chromium plated?
Are the proposed access points for maintenance adequate and sufficiently convenient?
Will there be any undesirable noise from down-pipes, for example?
Is the proposed cold water storage capacity adequate?
Are there any locations where possible pipe leakage would be highly undesirable? Should re-routing be considered to avoid the risk?
Are there any special requirements, and are these satisfied, e.g. precautions against legionnaires' disease?
Has the contractor been made aware of any employer specific requirements for water services, and have the requirements been met?

(xv) Heating, ventilation and air conditioning

Is the onus entirely upon the contractor to achieve specific performance requirements in all rooms and areas?
Is the risk for setting the performance criteria with the contractor or the employer?
Are the performance criteria for temperatures and air changes in accordance with the recommendations of the CIBSE guides, if not are there good reasons for the variance?

Have predicted energy cost been calculated?

Are warm-up periods important? If so, are they predicted and satisfactory?

Is the zoning of all systems satisfactory from the point of view of flexibility economy and convenience?

Are the controls and energy management systems adequately described, and are they what is wanted?

Is the heat source, plant and equipment all described adequately and appropriate?

Has the contractor submitted schematic diagrams of all systems so that the philosophy can be understood?

Are the heating appliances suitable from the point of view of comfort appearance and any other special requirement (for example, low surface temperature radiators, safety, tamper-proof controls, etc.)?

Are air speeds in ducts, and from ventilation grills within desirable limits?

Are there any peculiarities, experimental or innovative aspects to the concept design which are unproven and therefore risky?

Does the systems rely on occupational patterns which may change in the future? For example, is there a heat-pump system which draws heat from one part of the building to heat another?

Is the contractor aware of all parameters regarding noise from the building and its plant? For example nuisance could be caused to nearby residents by noise from plant operating at night.

Has the contractor been informed of the characteristics of any of the employer's plant or equipment which could affect his systems?

Are there any other special requirements, and have they been fulfilled?

Has the contractor included for all costs in connection with incoming supplies?

(xvi) Fire fighting installations

Have the insurers been consulted, and have their requirements been conveyed to the contractor and included in the proposals?

For sprinkler systems, has the adequacy of the mains water supply been checked, is it adequate and if not has the contractor included adequate storage facilities within the proposals?

Has the contractor included for loose items of fire fighting equipment?

Will the contractor be responsible for meeting the local fire officer's requirements?

(xvii) Electrical systems

Has the contractor included all costs in connection with the incoming supply, including a sub-station if necessary? If a PC or provisional sum has been allowed, is it realistic, and is the contractor prepared to take the risk by making the prices firm?

Has the contractor estimated the overall power supply requirements? Are the calculations satisfactory, and do they take into account any special needs of the employer? Are the diversity factors reasonable?

Are the distribution boards suitably located?

Are the main cable routes identified, and are they satisfactory?

Has the contractor included schematic diagrams for all electrical systems, including lighting, power, alarms, communications, computer and IT, public address, listening and viewing, stand-by generation and un-interrupted power, etc.? If not, are they adequately described?

Has the contractor estimated the total power consumption/costs?

Are lighting levels specified in all rooms and are they in accordance with the recommendations of the CIBSE guide? If not is there good reason for the variance?

Are the lighting levels specified at or on the appropriate plane, and with appropriate limits on the variation? Note, for example, sloping drawing boards.

Has the contractor assumed, correctly, or otherwise, that the employer will install his own task lighting in conjunction with his furniture or equipment?

Are there any special requirements for types of light fittings, and have they been met in the proposals? For example, colour characteristics.

Is the zoning of the light switching described, and is it satisfactory?

In areas such as offices, does the proposed lighting layout and switching provide adequately for future flexibility? On the other hand is it over-complicated, and therefore, unnecessarily expensive?

Has decorative lighting been adequately described, and is it satisfactory? If a PC sum has been included, is it realistic?

Are the small-power supply sockets identifiable, and are they adequate for the employer's envisaged needs, including a degree of flexibility?

Are the supplies for the employer's plant and equipment adequately defined, and are they of suitable capacity, and in the correct locations?

Is the interface between contractor's supply and the employer's plant and equipment precisely defined?

For stand-by generation and uninterrupted power supply installations, is the performance specification clear?

Are all the secondary systems clearly described, and in accordance with the requirements of the employer? Has the contractor included a lightning protection system in accordance with the relevant British Standard, or checked that it will not be necessary?

(xviii) Lifts, hoists and escalators

Are all lifts, hoists and escalators clearly specified in terms of speed, sizes, opening sizes, load capacities, operational characteristics, and finishes?

(xix) External Works

Has the contractor included any 'assumptions' that could lead to the possibility of additional cost to the employer?

Is it possible to identify from the contractor's proposals, what surfacing or treatment he will apply to all external areas, or are there any areas upon which he is 'silent'?

In the case of hard surfacing, has the contractor specified any limits as to the use of any areas, and is he aware of the use to which the employer intends to put to roads, hardstandings and parking areas?

Are there any areas which will be subjected to unusually hard wear, or where damaging substances may be spilled?

Is the employer satisfied with the appearance of the proposed surfacing?

Has the contractor taken fully into account the requirements for all vehicles, for such things as: deliveries, maintenance, car and lorry parking, fire appliances, etc.?

Is the pedestrian access suitable, and where necessary, segregated from vehicles?

Is access for disabled persons satisfactory?

Is the soft landscaping fully described?

Is the maturity of the proposed shrubs and trees specified, and is it satisfactory?

What responsibility will the contractor have for maintenance of the planting during and after the defects liability period?

Are any existing trees to be preserved, and what are the consequences if the contractor damages any such trees?

Has the contractor included for all external structures, buildings, features, barriers, fences and the like?

Are they adequately described?

Has the contractor included for all necessary works outside the site boundaries; for example, pavement cross-overs and all consequential works, such as the lowering of service mains?

(xx) Drainage

Has the contractor included for all costs in connection with the drainage systems, including all local authority requirements, such as: separation of systems, petrol interceptors, balancing tanks for surface water, etc.?

Are there any provisos or qualifications which could affect the employer's costs; for example, assumptions as to the adequacy and depth of existing outfalls?

Has the contractor included any statement as to the intensity of storm which the surface water drainage system is designed to accommodate, and what is the likelihood of flooding, albeit temporary?

FOLLOW-UP AFTER CHECKING THE PROPOSAL DOCUMENTS

Just as no contractor is bound to submit his proposals in accordance with the requirements of the enquiry, neither is the employer bound to accept proposals in the form the contractor gives them. Thus on receipt of the tenders and after a thorough check of the contractor's proposals the employer may seek answers to the questions which will have inevitably arisen.

Follow-up for Employer-led-Design Projects

With employer-led-design tenders, it is usually possible to prepare a questionnaire for all the contractors which seeks confirmation of relevant points in order to 'equalise' the basis of the tenders.

The types of questions which could be put to the contractors include, for example:

Problem: contractors differ in the way they describe their conditions for acceptance of the findings of the soil survey. One accepts all risks, others do not.

Possible question to all the contractors: have you relied upon the findings of the soil survey issued with the enquiry, and what adjustment, if any, would you make to your tender sum if you were obliged to accept all risks in connection to the ground conditions?

Problem: some contractors have specified particular lighting levels and air change rates in various rooms and they vary from contractor to contractor.

Possible question to all the contractors: have you complied in every respect with the recommendations of the current CIBSE guide? If not where have you varied, and what affect will it have on you tender sum to bring your proposals into line with the recommendations?

Problem: one contractor states that he has contacted the local electricity board, and they suggest verbally that a new sub-station will be necessary. Other contractors, either include a provisional sum, or say that in their view a sub-station will probably not be required.

This problem of comparison may be resolved by instructing the contractors to include a specific provisional sum for the cost of the incoming electrical supply, and to make the necessary adjustment to his tender sum.

Follow-up for Contractor-led-Design Projects

The employer has more difficulty with contractor-led-design proposals, because many of the questions to raise could apply to only one of the contractor's proposals, and so the very raising of the question itself to other contractor's would give them information which should remain confidential.

That is, it is unfair to a contractor with original ideas to divulge the idea to others so that they may gain advantage from it.

The employer may seek the answers informally through discussion with one or more contractors, or if he is concerned to demonstrate absolute impartiality he will have to adopt a more formal basis.

First he may issue a questionnaire, in the same format as described above for employer-led-design proposals, for those points which are common to all contractors.

Secondly he may issue particular questionnaires to contractors for the points which are not common to all. His objectives are:

(i) to seek clarification of any aspects of the proposals that are not clear,
(ii) to find out the effect on the contractor's tender of any changes that the employer requires of a contractual or commercial nature,
(iii) to ascertain the effect on the contractor's tender of any design changes that the employer may decide are necessary for his requirements to be met,
(iv) to ascertain the effect of any alternatives that either the employer or the contractor may propose that are potentially beneficial.

THE EMPLOYER'S FINAL CHOICE

Upon receipt of the answers to all his questions from the contractor(s), the employer will have all the facts at his disposal upon which to select the contractor for the project.

In theory it is difficult to make the choice where the proposals are disparate, as they could well be in contractor-led-design proposals, but in practice the choice is usually not so difficult. A parallel may be drawn with the process that a prospective purchaser goes through before he buys a car.

The purchaser considers for example, appearance, space for passengers and luggage, power for acceleration and speed, servicing costs, cost and availability of replacement parts, insurance costs, fuel consumption under various conditions, depreciation and all the various gadgets and extras which may, or may not, be available on the models which are under consideration. Most people would agree that they would be in two (or more) minds faced with such decisions, but few would say that they are incapable of making a reasoned choice.

The same would be true of a prospective house purchaser, and in his case whatever choice he makes will stay with him for a long time.

Thus the employer will identify those characteristics which are most important to him and he will give weight to them accordingly.

The lowest price may not be the final and most important deciding factor. It has been said that 'What good is a bargain, if it is not what you really want?', and so it is with building projects: employers are well advised to make sure that they know what they are getting for their money; as far as they can reasonably tell, ensure that it is what they really want; if they themselves are not capable of making valued technical judgements ensure that they are well advised; and above all it is suggested that an employer gives most weight to the contractor who gives him the most confidence that he understands the special requirements of the project and that he has the experience, capability and resources to fulfil his obligations.

A large air-conditioned office building in London Docklands, designed in detail by architects employed by the client. The development was sold-on and the new client changed the procurement route to 'Develop and Construct', instructing that the design team would be transferred to the contractor under a novation agreement. The JCT 1981 form of contract was used.

Building owner: Edgegrey Thames Quay Ltd
Design—Build contractor: Tarmac Construction Ltd

(Photograph: Stuart Galloway)

A primary school of traditional materials set in wooded surroundings in Reading. The building was procured using the 'contractor-led-design' Design–Build (two-stage tender) route and the JCT 1981 form of contract.

Client: Sisters of St Mary Madeleine
Design–Build contractor: Tarmac Construction Ltd
Contractor's architect: Kenyon McCrory

(Photograph: Graham Jackson)

A mixed retail and commercial city centre development in Coventry. It was originally tendered as a 'Develop and Construct' BPF/ACA contract, then a second stage was introduced to seek economies. The contractor's architect redesigned parts of the building to produce the elevations illustrated and the fabric canopy which shelters the Lady Godiva statue.

Client: St Martin's Property Corporation
Design—Build contractor: Tarmac Construction Ltd

An office building with basement car parking in Brentwood, Essex, built of traditional materials to harmonise with the surroundings. A 'contractor-led-design' negotiated JCT 1981 contract.

Client: Urban Land Properties Ltd
Design—Build contractor: Tarmac Construction Ltd
Contractor's architect: Ellis Williams Partnership

(Photograph: Key Photo)

III The Contractor's View

11 Contractor's Responsibilities/Risks

'Do the duty nearest thee, which thou knowest to be a duty
Thy second duty will already have become clearer'

Byron [*Satori Resartus*]

KEY TOPICS

* Additional obligations of the Design–Build contractor
* Responsibilities and risks in relation to:

  Design
  Appointment of consultants
  Specialist subcontractors
  Pricing

INTRODUCTION

This chapter is headed thus because any additional obligation, duty or responsibility, which a contractor takes on, constitutes an additional 'risk' to him. Self evidently, the Design–Build contractor assumes the same responsibility for the construction work as he does in a Traditional Contract, and as the name implies, he also accepts responsibility for that part of the design which he or his consultants undertake. However, there is a lot more to 'design' than preparing working drawings and ensuring that they comply with the requirements of the contract.

ADDITIONAL RESPONSIBILITIES FOR THE DESIGN–BUILD CONTRACTOR

A Design–Build contractor's areas of additional responsibility can be divided, and subdivided, as follows:

Matters Related to Design

   (i)   Function
  (ii)   Appearance
 (iii)   Durability
 (iv)   Compliance with Employer's Requirements, if JCT 1981 Form
  (v)   Compliance with Contractor's Proposals, if JCT 1981 Form
 (vi)   Compliance with statutory requirements

(vii) Accuracy
(viii) Buildability
(ix) Coordination
(x)   Timing

Appointment of Consultants (if external)

(i)   Choice
(ii) Terms
(iii) Services
(iv) Professional indemnity insurance

Specialist Subcontractors

(i)   Design
(ii) Tenders
(iii) Programme
(iv) 'Builder's work in connection'

Pricing

(i)   Quantities
(ii) Description
(iii) Coordination between elements
(iv) Elimination of omissions and gaps
(v) Fees

These responsibilities are discussed, in turn, through this chapter.

RESPONSIBILITIES/RISKS RELATED TO DESIGN

Function

The contract terms will govern the extent of the contractor's responsibility for the 'functioning' of the completed building, and its fitness for the purpose for which it is intended.

As we have seen in Chapter 4, a 'fit for purpose' obligation, included in contract terms and agreements with consultants, has legal significance.

Usually, it is the employer who defines the purpose of the building and its constituent parts, and if he does not do so in the enquiry documents, the contractor is well advised to establish the purpose(s), otherwise he cannot design the building so that it will function properly. If the employer is unable or unwilling to identify the intended purpose of the building and all of its rooms and spaces, the contractor should make it clear either in his tender or in the contract documents, or in writing at an appropriate time, that his design liability is qualified accordingly.

The importance of establishing the use of rooms or spaces can be seen from these examples:

Example 1
The purpose of an inter-connecting room adjacent to the Chairman's office in an administration building may be unspecified. The building may rely on opening windows and natural ventilation generally. The enquiry documents may stipulate that mechanical ventilation is to be included where necessary.
If the room is to be a conference room it could well need mechanical ventilation, depending upon its capacity, the size of the opening windows and other factors.
Whereas, if it is a waiting room for the Chairman's visitors then natural ventilation may well be adequate.

Example 2
An employer enters into contract for the design and construction of a 'home for the elderly'.
He does not make it clear that the home is to be used as a nursing home, and as such it will have to be registered by the local health authority before it may be used for that purpose.
Health authorities lay down strict standards which must be met for registration, and these requirements can add considerable costs, in particular, to the heating and hot water systems.

Both these examples can have similar consequences, in that the contractor is in danger of either under-providing, with consequential risk, or over-providing with the consequence of either being uncompetitive or incurring unnecessary costs.

There are many commonly described rooms or spaces where their use is not entirely implicit in the description, and where differing uses could have a significant affect on the design and servicing. For example: entrance halls, atria, food preparation rooms (extent of cooking), office/meeting rooms, drawing offices – lighting, any rooms containing employer's specialist equipment, rooms containing chemicals etc. (what chemicals?), warehouse/stores, (what goods?), etc.

Design for Appearance

The enquiry documents may or may not include the design of the exterior (and interior) of the building. However, it is one thing to specify materials and illustrate their use on 1:100 drawings, but it is entirely another to detail the angles, corners, junctions, etc., all of which could have a marked effect on the building's appearance and quality. Transoms and mullions may be shown on the enquiry drawings as narrow single thickness lines, but the actual thickness of the members could vary considerably, so spoiling, or enhancing, the style of the building.

So, even where the enquiry specifies the 'appearance', there is still a lot for the contractor's design team to do to turn these ideas into 'reality'.

The use of high quality or expensive materials will not automatically guarantee a satisfactory appearance.

In their tender documentation, contractors are well advised to be as specific as they can be on this subject. If they 'cut corners' in the detailing to keep the price low, they may well end up having to spend extra money bringing the work up to standard and, on the other hand, if a contractor prices a 'Rolls-Royce' finish and detailing, he may price himself out of the job unwittingly.

Design for Durability

No matter what the terms of a particular contract are, it is the contractor's responsibility to ensure that the durability of the building and its parts meet the requirements of the contract in the broadest sense; and where the contract is silent, the contractor must ensure that the standards are appropriate.

To achieve the required standards contractors must consider the required life expectancy of the building and, against that, specify the fabric and finishings of the building taking into account wear and tear due to use, and deterioration due to environmental effects.

Design for Compliance with Employer's Requirements

Contract Drawings and Specification

It is the contractor's responsibility to ensure that his tender complies with the employer's requirements as expressed in the enquiry, and if it does not comply in any respect, he must safeguard his position by pointing this out in his tender.

During the period between the tender and completion of the contract documents, changes often take place which affect the employer's requirements compared with the way in which they were expressed in the original enquiry. Contractors must ensure that their proposals are brought up-to-date to comply with the employer's requirements as expressed in the contract documents, and that the contract sum also reflects any such changes.

Working Drawings and Specification

The contractor must also ensure that the design, working drawings, detailing, specification, and specialist subcontractors' designs as they develop remain consistent with the employer's requirements, as expressed in the contract, with obvious potential cost penalties if they do not.

The employer may have included a 'sanctioning' process into the contract procedures. It is standard in the BPF System. Employer will generally ensure that any such sanction does not relieve the contractor of his design responsibilities under the contract.

Materials to the Employer's Approval

Sanctioning processes are to be distinguished from any stipulation in the contract documents that certain items are to be to the 'Employer's approval', e.g. finishing materials, fittings, etc.

If the documents contain this provision contractors are at risk for additional costs if they do not ensure that the contract includes a mechanism for controlling the costs involved, or for limiting the employer's choice. This may be done by establishing a base cost by way of a provisional or prime cost sum in the contract, but the contractor must ensure that the terms of the contract provide for the resolution of such sums. For example, the JCT 1981 Form only allows the Contract Sum to be adjusted in relation to provisional sums if they are included in the Employer's Requirements.

Another way of controlling the cost involved in items which are to the employer's approval, or are to his choice, is to ensure that the contract documents stipulate that the choice is to be from a particular manufacturer's range of products or colour for example.

Design Compliant with Contractor's Proposals

We have seen that contractors have a responsibility to ensure that the design as it develops complies with the requirements of the contract; from the contractual point of view, this is a minimum requirement because they also have to comply with their other obligations. However, if they exceed that which they have allowed for in their proposals, they will incur unrecoverable costs.

So, a contractor must first ensure that the contract sum is calculated taking into account any changes to his proposals which have taken place after submitting his tender and prior to signing the contract; and secondly he must ensure that the design, working drawings, schedules, and specialist sub-contractors' works do not gratuitously exceed the provisions made in his proposals.

Contractors must ensure that their checking systems and lines of communication with their designers are such that they do not incorporate anything within their developed design that is likely to involve the contractor in additional costs, either unknowingly or unnecessarily. With the speed with which most Design–Build contracts proceed through design to construction, this can be an exceedingly difficult task. To control the design in this respect requires experienced personnel to exercise the necessary disciplines and procedures. It may assist contractors to include a clause in the consultants' appointments to the effect that the consultant is to report to the contractor any likely increases in cost through design development.

Design to Comply with Statutory Requirements

The terms of the contract will determine the extent of the contractor's responsibility for obtaining statutory approvals and for ensuring that the design complies with all relevant regulations. It will also dictate whether the contractor would be entitled to reimbursement in the event that such compliance involves him in additional expense. Individual contracts may well differ in this respect.

In essence, the standard forms deal with the subject as follows:

JCT 1981 Form – Statutory Requirements (Clause 6)

In principle, Clause 6 of the JCT 1981 Form places the onus upon the Contractor to ensure that the Works comply with Statutory Requirements and the Contract Sum will not be altered unless either:

(i)   an Employer's 'change' brings about a 'contravention'. In which case resultant additional cost will be reimbursed to the Contractor; or

(ii) if the Employer's Requirements specifically state that they comply with Statutory Requirements and this transpires to be incorrect; or

(iii) if there is a change in Statutory Requirements after the Base Date named in the Contract.

Example of (i):

The Employer changes part of an office from an open-plan to a cellular layout. If this affected the means of escape, the evaluation of the change would include all costs associated with correcting that situation, e.g., extra fire doors, or additional escape stairs, etc.

Example of (ii):

In practice, it is rare for enquiry documents to state specifically that the Works, in whole or in part, comply with any particular statutory require-ment. Indeed, it may well be considered unwise for an Employer to take on this responsibility. However, there are circumstances when the Employer is unable to furnish the Contractor with sufficient information (e.g. occupancy levels); or where the Employer himself may have special expertise (e.g. waste processing) and it may be most expedient for him in such circumstances to take on the responsibility for ensuring that particular parts or aspects of the Works comply with particular relevant statutory requirements. In either case the Contractor should ensure that any such provision is incorporated into the Employer's Requirements, whether or not it was included in the enquiry documentation.

Another example is when the Employer makes application himself for a statutory approval prior to the appointment of the Contractor.

BPF Form – Statutory Requirements (Clause 1.6)

The provisions of the BPF Form are different from the JCT 81 Form.

In principle the Contractor may assume in his tender that the elements of design undertaken by the Client comply with Statutory Requirements and if they do not then the Contractor will be paid the cost of any 'change'. That is, unless the Contractor could have reasonably foreseen that the Statutory Requirements would not be met, in which case any change would be at the Contractor's own expense.

In the BPF Form, the Contractor is responsible for ensuring that his own working drawings and details comply with Statutory Requirements and he will not be paid any *abortive* costs if they do not comply; or if a 'non-compliance' is carried forward from the Client's design into his own working drawings or details.

Difficulty in Determining Responsibility for Compliance with Regulations
Circumstances which make the question of liability for compliance with statutory requirements uncertain, can arise in a number of ways. Especially, if for one reason or another, the contract excludes certain works which form an integral part of the completed building, and without which the building would be incomplete.

Common examples of this are:

(i) Pre-contract works by others.
   Pre-contract enabling works, or works constructed by others prior to the commencement of the Design–Build contract.

(ii) Fitting-out works by others.
   There are many ways in which fitting-out can affect compliance with statutory regulations. For example, loading of the structure, means of escape, provision of fire-fighting appliances, ventilation of rooms, etc.

(iii) Alterations resulting from a change in intended use.
   Many regulations relate to the specific use of the building. We have seen earlier in this chapter, the importance to both employers and contractors of knowing the intended use of the building and its parts. However, a property developer, having a 'speculative' office or industrial development built, may specify the intended use, but a purchaser or tenant may then wish to put the building to different use. In such a cases contractors are unlikely to be responsible for any consequential additional costs, unless the particular contract terms are so worded that the contractor could be made liable.

(iv) Works which result from dividing a building into different tenancies.
   A speculative development may be built under a contract which requires the contractor to design the building such that it may be divided into separate sections for sale or lease to different companies. The precise terms of the contract will determine the contractor's responsibility, but the contractor must be wary of accepting blanket requirements for flexibility in the way the building may be divided. Fire resistance of building elements and means of escape are two factors which can be affected quite significantly by the position of notional boundaries between different tenants or owners.

Contractor's Responsibility for Accuracy in the Design

Contractors' working drawings and schedules must be accurate in every sense of the word. They have no claim against the employer for any inaccuracies in such documents. This represents a very real 'risk' to contractors, compared with a Traditional Contract, where problems of this type so often lead to delay, disruption, loss and expense for which the contractor would be entitled to reimbursement and an extension of time.

The terms and circumstances of a particular contract will determine responsibilities in respect of the accuracy of information provided by the employer.

Frequently the employer will furnish tendering contractors with survey information including ground exploration. The employer has the choice whether to stipulate that such information is provided in 'good faith', and that the contractor will have no claim should the information subsequently prove inaccurate or, in the case of soil surveys, untypical. But on the other hand, the contractor has the choice whether he will accept such terms. He may either decline to tender or qualify his tender. If he accepts the terms, he should evaluate the risk and include a contingency.

It may seem to be to the employer's advantage to place this risk with the contractor. However, the employer will 'pay' whatever contingency the successful contractor includes in his tender, whether the contractor incurs the cost or not; and so, in some cases the employer may judge it prudent not to pass on all these risks to the contractor, but instead make some contingency of his own for use only if such a problem does arise.

Contractor's Responsibility for Buildable Design

Contractors have every opportunity to ensure that drawings and details produced under their direction are 'buildable'. In fact this is perceived to be one of the main advantages of Design–Build contracting.

However, there is an element of risk to the contractor. If his designers do not perform well and he, the contractor, does not manage them well, he could find himself with unworkable details, with no one to blame but himself.

It is not just a question of seeing that the details are not unbuildable, but the real skill comes in making sure that the design allows the work to be built as easily as is practical (without detriment to the intent of the design), because this not only leads to economy in construction, but undoubtedly, the easier anything is to build, the better will be the quality of workmanship.

To achieve buildability requires a disciplined approach on the part of the contractor. First he is well advised to hold discussions between the designers and construction staff at early and other appropriate occasions to discuss building methods, availability of materials, etc.; and secondly he must build into his programme and procedures the provision for the construction staff to have the opportunity to examine the designers' working details in sufficient time to allow the designers to incorporate any agreed improvement.

Contractor's Responsibility to Produce Coordinated Design

First, it is appropriate to differentiate between 'project coordination' and 'design coordination'.

In Traditional Contracting, architects are the design coordinators and design team leaders. It can be dangerous to depart from this in Design–Build, because architects, and not contractors' management staff, have the training and experience to fulfil this function.

To 'coordinate' the design means to ensure that every facet or element of the design and its parameters are compatible.

That is to say that the designers, which includes the consultants and any of the subcontractors who have design responsibility, all work to the same parameters and objectives, and that each is aware of the other's design and specifications so that each may take due account of all relevant information in carrying out his own design and specification.

For example, to ensure that a subcontractor's design is fully coordinated into the overall design, involves checking their drawings and specifications for tolerances, dimensional coronation, coordination of fixings, and that the finish, appearance, durability and performance criteria are all compatible with other elements and the overall intent of the design.

Discharging this responsibility, in the Design–Build situation, is difficult for the architect because he is often remote from the designer subcontractors and he has no direct control over the flow of information to and from them. To make matters worse, these subcontractors are often appointed at a relatively late stage by contractors for commercial reasons.

At the outset, contractors must plan and programme the design work with their consultants taking due account the need for coordination of the work of all parties involved, and failure to do so constitutes a very real risk to the contractors.

Many contractors, especially those who employ independent consultants, use the title 'Project Coordinator' for the member of their staff who interfaces between the consultants and the construction staff. Often, he is also the point of contact with the employer's representative. He is not the 'design coordinator'. This is true except when a contractor defines the duties differently for his own staff.

Contractor's Responsibility to Produce Information on Time

Unless, in a particular contract, the contractor is dependent upon information from the employer, and that information is issued late, contractors have no redress if the design and working details are not issued to suit the construction programme.

Contractors must consider lead-in times and sanction requirements, as well as procurement and construction periods when preparing tender and construction programmes.

If a contractor allows insufficient time, or his design team fall behind the design programme, the contractor is at risk for the cost to himself of delay (or acceleration) and, quite possibly, the cost of liquidated and ascertained damages. This is a risk that the Traditional Contractor does not have.

APPOINTMENT OF CONSULTANTS – CONTRACTOR'S RISK/ RESPONSIBILITY

Currently the contractors' design for the majority of Design–Build contracts is undertaken by independent consultants employed by contractors on a one-off, ad-hoc basis and the first risk for a contractor is in his choice of consultants.

The second is the terms of the appointments. Even in those situations where the consultants were first employed by the employer for pre-tender work, and then by the contractor, under a novation agreement for the post-contract work, the contractor still remains responsible for the terms of the appointments.

The risks that a Design–Build contractor has in making the appointments can be summarised briefly, as follows:

(i) Failure to make adequate provision in the tender for fees and disbursements.
(ii) Failure to ensure that the consultant includes the costs for all the services required by the contractor, whether a contractual requirement or not.
(iii) Having a 'gap' between the services to be provided by the consultants.
(iv) Failure to recognise the need for specialist consultant's services.
(v) Differences in design liability between the terms of the main contract and consultants' agreements.
(vi) Agreeing, under the terms of the main contract, to provide collateral warranties from the consultants in favour of the employer or others, and subsequently finding that the consultants are not prepared to enter into such agreements.
(vii) Failure to establish a satisfactory agreement in respect of timing for the production of information.
(viii) Failure to ensure that the consultants have adequate professional indemnity insurance cover.

SPECIALIST SUBCONTRACTORS – CONTRACTOR'S RISK/ RESPONSIBILITY

Specialist Subcontractor's Design

In Traditional Contracting specialist subcontractors are usually nominated, or at least 'named' by the architect. Usually, contractors are careful to ensure that they avoid any responsibility for the design, and so it is quite usual for subcontractors to enter into a form of warranty for their design in favour of employers.

In Design–Build, contractors assume full responsibility for subcontractors' performance, including design, hence contractors are at risk if there are any deficiencies in the design; their only recourse is to the offending subcontractors.

Specialist Subcontractor's Tenders

In network planning terminology, specialist subcontractors are invariably on the 'critical path' in the tender production programme.

If a Design–Build contractor fails to secure tenders from major specialist subcontractors, before the deadline for submitting his own tender, he has only three options, namely:

(i) He may decline to submit his tender if the subcontract work concerned constitutes a significant part of the project.
   In which case, he has wasted the time and cost in tendering; and he risks spoiling his relationship with the employer and his representatives, or
(ii) He may submit his tender, including a provisional sum for the subcontractor's work, but he is unlikely to secure the contract, or to please the employer, or
(iii) He may make his own assessment of the allowance to include, and the risks involved in doing this are obvious.

Specialist Subcontractors' Programmes

In Traditional Contracting contractors may reject a nomination if the subcontractor cannot agree to the contractor's programme requirements. In Design–Build, a contractor has no such safeguard.

Builder's Work in Connection with Subcontractors' Works

In Traditional Contracting, the builder's work in connection with the subcontract works is measured and valued according to the rates in the contract bills.

In Design–Build, contractors must make provision in their tender for this work at a time when it is rarely accurately quantifiable. Contractors risk over-pricing the work and hence becoming uncompetitive, or under-pricing it, and thereby incurring losses.

CONTRACTOR'S PRICING RISKS

Under the previous headings, reference has been made to risks in 'pricing' the tender. Here, we look at these risks in more detail.

In essence Design–Build contracts are lump-sum-fixed-price contracts, not subject to measurement in any way, and with a few exceptions, contractors are obliged to complete the Works described in the Contract Documents for the Contract Sum. So the Contractor takes responsibility for the accuracy of the Contract Sum.

The aspects of a Design–Build estimate which vary from Traditional Contracting include the following:

(i)   Quantities in the take-off
(ii) Description of the elements in the take-off
(iii) Coordination between elements
(iv) Elimination of omissions
(v) Estimate of fees and other particular Design–Build costs.

Pricing Risk – Quantities

In Traditional Contracting the quantities for the bill of quantities are measured by the employer's quantity surveyor, and even if the bill of quantities is not described as 'approximate', any innaccuracy is at the employer's risk.

In contrast, the Design–Build contractor is entirely at risk for the quantities which he measures.

Pricing Risk – Descriptions

At first sight it may appear superfluous to include 'description' as a pricing risk in addition to 'quantities'. However, it is worthy of separate mention because in Traditional Contracting a standard method of measurement is used which defines the way in which elements of the work shall be measured and described, and competing contractors and subcontractors all price the work with a common understanding.

In Design–Build, the contractor is responsible for the description of the elements in his take-off. Whilst the contractor may seek to prepare the take-off in accordance with a standard method of measurement, he may have insufficient time to do so. He then resorts to non-standard descriptions which involve him in risks in relation to subcontractors' prices, and also in his own pricing where the take-off is done by one party and the pricing is by another.

Pricing Risk – Coordination between Elements

In pricing works, contractors are at risk if any elements of the tender specification or design are not mutually compatible. A simple example demonstrates the point, and the risk:

Suppose, in a block of flats, the contractor's designers specify a particular type of precast flooring, without a structural topping, supported on external loadbearing blockwork and corridor walls. Cross walls are to provide lateral restraint to the load-bearing walls, but this requirement is not marked on the drawings.

Suppose the contractor, in finalising his tender, prices stud partitioning cross walls in lieu of blockwork and plaster, believing that they performed no structural function, and accordingly cuts time and money from his tender.

The resultant building would have been structurally unsound, and the contractor would have the option of reinstating the blockwork, or introducing a different type of floor which could act as a 'plate' to provide the lateral restraint required. In either case he would have incurred additional cost and, possibly, a time penalty.

Pricing Risk – Elimination of Omissions and Gaps

Errors in quantities in the take-off was the subject of a previous paragraph, and they occur simply through miscalculation on the part of the taker-off. Omissions and gaps occur for different reasons. For example:

(i) Failure on the part of the design team to identify the need for a particular element, e.g.,

the need for an escape staircase,

fire barriers above a ceiling,

mechanical ventilation for a particular room or lobby.

(ii) Gaps in the estimate caused through lack of coordination between sections in the take-off, e.g.,

ventilation grills not included in either the mechanical section, nor the builder's work section,

support for light fittings not included in either the electrical or the suspended ceiling sections,

embedded fixings neither measured in the structural work, nor in, for example, the curtain walling section.

The most common and most expensive errors in Design–Build contractors' estimates are the complete omission of items of work, and it is in the nature of things that one is more likely to overlook a requirement than to include unnecessary items, and so contractors do not normally even have the benefits of 'swings and roundabouts'. The risk of loss is, therefore, considerable.

Pricing Risk – Fees and Other Particular Design–Build Costs

The Design–Build estimator is required to cover a number of items which his Traditional Contracting counterpart does not. The most significant of these are:

(i)   fees to consultants,
(ii) fees to the Local Authority,
(iii) contributions to Statutory Undertakers,
(iv) cost of mock-ups and tests,
(v)   cost of models, perspectives, etc.

12 Selection and Appointment of Contractor's Consultants

'You pays your money
and you takes your choice'

Punch

KEY TOPICS

- Novation agreements
- Selection of consultants
- Agreements
- Conditions of consultants' appointments

INTRODUCTION

In this chapter we consider contractors' selection and appointment of external consultants. The agreements may not be drawn up formally until, and if, the contractor secures the contract, but nevertheless, the terms of the appointment should be settled by the parties before the contractor submits his tender.

It is appropriate to identify five integral parts of an agreement between a consultant and his employer (in this case the contractor), they are:

(i) The Agreement itself
(ii) The Conditions
(iii) The definition of the scope of the works for which the consultant is to provide his services
(iv) The services that are to be provided
(v) The fees.

The various topics are discussed within different chapters, as follows:

131

Novation Agreements

What is a Novation Agreement?

Occasionally, an employer may stipulate that the successful contractor is to engage one or more of the employer's consultants to complete the design and detailing of the project in the post-contract stage. Such consultants are employed by contractors under 'novation agreements', i.e. a 'novation' agreement is a new agreement in substitution of a previous agreement between the consultant and the employer.

When is 'Novation' Practical?

It is not possible nor practical for employers to impose such a condition in all varieties of Design–Build, as Table 12.1 shows.

Table 12.1   *'Novation' of Consultants and the different varieties of Design–Build*

Variety of Design–Build	Comment
Develop and Construct	Most common variety where novation is applied
Design–Build (single-stage tender)	Novation often suggested by employer, but contractors should take care
Design–Build (two-stage tender) (Design competition)	By definition, novation is not possible
Negotiated Design–Build & Design and Manage	Possible, but in this case it would usually be a 'named' consultant, in that the employer would be unlikely to have already entered into an agreement with the consultant
Turnkey	Novation would not normally be applicable.

Potential Difficulties with 'Novated' Consultants

Before agreeing to accept the imposition of a 'novated' consultant, a contractor should go through the process of checking that he and the consultant will be able to work together fruitfully right through to the completion of the project, otherwise the contractor could be better off declining the opportunity, or seeking the employer's agreement to waive the condition; this may actually be in the employer's interests, in the long run.

We have seen, from Chapter 11, the responsibilities/risks that contractors take on in Design–Build, and so they must satisfy themselves that they will have satisfactory and confidential access to the consultant during the tender period, bearing in mind that other contractors will also be in a similar position.

More specifically, the contractor should seek answers to the following points, at least:

(i)　There will be many proposals and details that the contractor will want his designers to consider, and so the contractor must check on what arrangements have been made to allow time for such discussion.

(ii)　How will any innovative proposals be kept confidential?

(iii)　What will be the consultant's attitude to contractor's alternative proposals devised during the tender period, and will the consultant be prepared to work-up such ideas?

(iv)　Will the consultant be able to devote sufficient time with the contractor at the conclusion of the tender period for detailed checks for errors and omissions, and to assist with risk assessment?

If the contractor can satisfy himself on all these points then, in principle, there should be no impediment to accepting the novation condition.

Criteria for Selecting Consultants (When the Contractor has the Choice)

One of the most important decision for a Design–Build contractor to make when embarking on any particular project is: who shall undertake the design work on his behalf?

Different contractors, and indeed different personnel within the contractors' organisations, have different criteria for selecting consultants. The person responsible for the selection will have his own order of priorities, but the factors to be considered will normally include the following:

(i)　In pre-qualification situations, will the consultant be of positive benefit to the contractor through his experience, presentation skills, approach, etc?

(ii)　During the tender period, will the consultant provide a service which will give the contractor the best chance of securing the contract?

(iii)　If the contract is secured, will the consultant provide a satisfactory standard of service in the post-contract stages?

(iv)　Will the consultant accept the contractor's proposed terms and conditions of appointment?

(v)　What level of fees will the consultant charge, both during the tender period, and during the post-contract period?

(vi)　Is there an opportunity for future reciprocal business?

(vii)　Does the contractor owe the consultant 'a favour'?

However in practice, upon receipt of an enquiry, contractors usually have very little time in which to consider these questions in relation to a previously unknown consultant, still less to obtain competitive fee proposals from consultants, and so more often than not, contractors will choose either:

(i) a consultant with whom he has enjoyed a good working relationship previously, or

(ii) one who had already made successful representations to the contractor in readiness for a potential ensuing enquiry.

CONTRACTOR'S CONSULTANTS' AGREEMENTS

The actual form of agreement between a consultant and a contractor can be relatively short, and typically, it will contain the following details:

(i) The date of the agreement

(ii) The two parties to the agreement

(iii)The substance of the agreement, by reference to:
Definitions
The attached conditions
The services to be provided
The payment, or consideration.

In contrast, as we see later, the description of the conditions and services can be quite lengthy.

CONDITIONS OF CONTRACTOR'S CONSULTANTS' AGREEMENTS

There are wide divergences between different contractors' terms of appointment of consultants. Some, particularly smaller contractors, hopefully not through ignorance, appoint consultants by way of a simple letter confirming the agreed basis for the fee calculation, but make little or no reference to the services required, nor the terms and conditions of the appointment. Others make reference to standard terms issued by the respective professional institutes, and others have their own company standard forms of agreement.

There are fundamental changes that need to be made to the RIBA and ACE Conditions, to adapt them for use between a contractor and his consultants in Design–Build projects, for example:

(i) Who is the employer, and how should the conditions be adapted to differentiate between the Employer in the Contract, and the consultant's employer, namely the contractor.

(ii) The whole question of nominated subcontractors is not applicable, but what is the relationship between consultants and the specialist sub-contractors appointed by the contractor?

For this reason, most experienced Design–Build contractors prepare their own standard forms of appointment. It is not our purpose, here, to suggest a model set of conditions, as this is a matter for each contractor to decide, but typically, contractors' agreements contain provisions for the following subjects. They are essentially the same for both architects and engineers:

(i)   Definition of the scope of the Works, for which the Consultant is to provide his services, normally by reference to an Appendix (see next chapter).

(ii) Reference to an Appendix which defines the Services which are to be provided (see next chapter).

(iii) The Consultant's obligation to provide the Contractor with information in a manner, and at a time to suit the Contractor.

(iv) The Contractor's obligation to provide the Consultant with information, which he needs, 'on time'.

(v)   The power, or limitations on the power, of the Consultant to issue instructions.

(vi) The Consultant's responsibilities with respect to cost control and monitoring.

(vii) The Consultant's duty of care; the level of duty, and to whom?

(viii)The Consultant's relationship with other consultants and specialist subcontractors.

(ix) The Consultant's obligations in respect of professional indemnity insurance, both during the Contract Period, and for the period of potential liability.

(x)   Conditions of copyright.

(xi) The duration of the Agreement.

(xii) Conditions in respect of termination of the Agreement, i.e. the circumstances under which it could arise; and the payment terms should it arise.

To complete an agreement between a contractor and a consultant it is necessary to define the scope of the works, the services and the fees payable. Each of these subjects is covered in Chapters 13 and 16.

CONDITIONS OF ENGAGEMENT OF CONTRACTOR'S QUANTITY SURVEYOR

Often contractors engage outside professional quantity surveyors to prepare bills of quantities to assist the contractor in the preparation of his tender. These topics are covered in Chapter 14 , 'Tendering'.

13 Contractor's Consultants' Pre-Contract Services

'Small service is true service, while it lasts'

Wordsworth [*To a child, written in her album*]

KEY TOPICS

- Architect's and Engineer's services
  Pre-qualification stage
  Tender stage
  Negotiation stage
- Scope of engineering works
  Structural
  Building services

INTRODUCTION

In this chapter, we consider the services that a contractor may require of consultants during the pre-contract period. It applies to projects where the contractor employs external consultants, but much of the contents could also be applied to the work required of in-house design departments. It may well be that contractors using in-house design would enjoy an improved performance if they were to use the same techniques and disciplines as they would with external consultants.

In the previous chapter we considered the Agreement and the conditions of the appointment by the contractor of his consultants, whereas in this chapter we consider two other elements, namely:

(i) The services that are to be provided by the consultants in the pre-contract stage
(ii) The definition of the scope of the works for which the consultant is to provide his services

Chapter 16 'Post-contract Services' completes the full picture, and it also contains comment on the fees for consultants' services

SERVICES REQUIRED OF THE CONSULTANTS

There are a maximum of three pre-contract work stages in Design–Build, none of which relates specifically to the 'Project Work Stages' as defined

136

either by the RIBA, BPF or the CAPRICODE procedures. Those project work stages are described in Chapter 7, 'Employer's Representation'.

The Design–Build contractor's pre-contract work stages are:

(i)   Pre-qualification stage
(ii) Tender Stage
(iii) Negotiation Stage

The services required of the consultants varies through the stages, and also for each of the design disciplines.

Pre-qualification Stage

All Disciplines

Not all projects have a pre-qualification stage, and even those that do, may not involve contractors' consultants. There will be situations, however, when an employer decides to go through a rigorous selection procedure, starting with all the contractors who are interested, and then successively eliminating them until he is satisfied that he has the best firms available to price his project. Not only could this process apply to the compilation of a list of contractors to tender, but also to the selection of a contractor for a negotiated contract.

For the consultants in a contractor's team, the work involved may vary from simply preparing a practice statement, to drawing up outline proposals of the proposed project to impress the employer with their technical expertise, flair and appreciation of the nature of the project. All projects are different and the contractor will lead the team and decide, in consultation with them, the level of effort required for any particular project. One thing they will bear in mind, is that the various parties should complement each other and where, say, a contractor has limited experience of a particular type of building, the architect, for example, may fill the gap making use of his experience.

Tender Stage, Architect's Services

At the start of the tender period, a contractor and his architect should have an understanding between them on the basis for the appointment so that each is clear on what is expected of the other.

The RIBA Appointment lists architects' services, but neither the services, as described, nor the stages within which they are listed, suit the Design–Build situation, and hence it is necessary to redefine them.

The following check list includes most of the services that a tendering Design–Build contractor would normally require:

Architect's Services, Tender Stage:

(i)   Examine all the available information for the project, and discuss with the Contractor the strategy for preparing the tender.

(ii) Visit the site, to ensure that all relevant restrictions, which may affect the design, are taken into account in the design and specification for the tender.

(iii) Check whether there could be any problem with noise or nuisance, which may have to be taken into account.

(iv) Advise whether, in the Architect's opinion, the Contractor should seek advice from any other consultants or specialists.

(v) Obtain all information with respect to the requirements of the local authority, if necessary, through consultation with them.

(vi) Advise on the scope of the work required in order to ensure that the tender may include all Works needed to make the Works complete and functional (unless specifically excluded from the tender).

(vii) Assist the Contractor in determining the need for specialist subcontractors, and where appropriate suggest the names of suitable firms.

(viii) Fulfil the role of Design Team Leader, and coordinate the design work of others into the scheme as a whole.

(ix) Advise the Contractor whether any elements of the Works should be subjected to testing to prove their adequacy prior to incorporation into the Works, so that the Contractor may make due allowance in his tender.

(x) Assist the Contractor in preparing a design programme, so that realistic allowances may be made for the periods for design and approvals.

(xi) Provide the Contractor with all necessary notes, sketches, specifications and drawings to enable the Works to be measured and priced accurately.

(xii) Provide the Contractor with drawings, technical specifications and the like, needed by the Contractor in order to seek tenders from specialist subcontractors.

(xiii) Advise the Contractor if there are any conditions or requirements contained within the enquiry which, in the Architect's opinion, would or could lead to difficulties, such that it would be advisable for the Contractor to limit or qualify his tender.

(xiv) Provide the Contractor with documentation to support his tender, in the form of drawings specifications, models, samples, etc. all as agreed with the contractor.

Even in employer-led-design projects, it would normally be necessary for a contractor to seek advice from his architect to advise on all the points covered by the check list, albeit that the work involved will be less than that involved in a contractor-led-design project.

Scope of the Works for the Architect's Services

Under normal circumstances, in both Traditional Contracting and Design-Build, architects are concerned with the design of the whole of the works.

In practice, those parts of the works which an architect is not to design are often left unspecified in both Traditional Contracting and Design-Build. This is an undesirable omission, in that without this understanding, it would be impossible to predetermine the amount of work required of the architect, and therefore what the agreed fee should cover.

In both the RIBA Appointment, and in the check list of services described earlier in this chapter, the onus is upon the architect to advise on the need for the appointment of other consultants, item (iv) in the list, and specialist subcontractors, item (vii). Thus, the architect may be deemed to be responsible for the design of the complete works, the extent of which he must specify (see item (vi) above); except for those elements covered by his recommendations under those items, (iv) and (vii).

Tender Stage, Structural Engineer

The ACE Conditions of Appointment for Engineers cover the services required of an engineer, but as with the architect, they are compiled to suit Traditional Contracting and not Design–Build. Hence the specific services required of an engineer must be considered afresh, as they are in the following check list:

Structural Engineer's Services, Tender Stage:

(i)   Examine all available information relating to the engineering works, and discuss with the Contractor and others, the strategy for the preparation of the tender with particular emphasis on the manner in which the engineering information will be presented to enable the Contractor to make an accurate estimate of the cost of the Works.

(ii)   Visit the site to examine and obtain any local information that is reasonably accessible, to assist in the design of the engineering works.

(iii)   Advise on the pre- or post-contract need for the Contractor to obtain or undertake any topographic surveys, geotechnical investigations, desk studies and the like to enable the Engineer to carry out his duties during the tender stage, or the post-contract stage.

(iv)   Visit and consult the local authority to ascertain any factors that may influence the engineering design, and the provisions which the Contractor should make in his tender.

(v)   Advise on the scope and nature of the structural/civil/drainage/traffic engineering works required in order to make the Works 'complete' and functional.

(vi)   Advise on the need to employ any other consultants in relation to the engineering works either during the tender period, or post-contract.

(vii)   Assist the Contractor in identifying the need for specialist subcontractors and suggest names of suitable firms; prepare the technical information required in order to seek quotations from such firms; assist in the evaluation of quotations received.

(viii) Provide the Architect with advice and information relating to the engineering works, so that he may coordinate the design development of the Works as a whole during the tender stage.

(ix) Advise the Contractor on the need for any tests, or the like, which may be required, so that the Contractor may make adequate provision in the tender.

(x) Advise the Contractor of any conditions or requirements contained within the enquiry which may lead to difficulties, such that it is advisable for the Contractor to limit or qualify his tender.

(xi) Assist the Contractor to prepare a design programme so that realistic allowances may be made for the times for design and approvals.

(xii) Assist the Contractor in assessing reasonable contingencies to be allowed for additional costs which may arise during the post-contract design development.

(xiii) Provide the Contractor with documentation to support his tender, in the form of drawings, specifications and the like, all as agreed with the Contractor.

The services described in the check list would normally be necessary for all the varieties of Design–Build, including even employer-led-design projects. The variables from project to project lie not in the description of the services, but in the work that may be involved in fulfilling each of the tasks. For example, engineers would have far less work in tenders for straightforward employer-led-design, simple buildings than they would in contractor-led-design, large or complex buildings. Nevertheless, unless the contractor is to be put at risk later, his engineer should go through all the steps in the list from (i) to (xiii).

Scope of the (Structural) Engineering Works

To avoid later difficulties, or misunderstandings, it is necessary to agree, at the outset, on which parts of the works the engineer is to concern himself with. The underground drainage, for example, is one element of the works which may be designed by the engineer, or by others; the engineer may not understand that the contractor expects him to check the stability of all internal block walls for example, and there are other such elements where there could be confusion.

It is more important in Design–Build than in Traditional Contracting to predetermine the scope of the engineering works, because in Traditional Contracting engineers' fees are normally calculated upon completion, and the percentage fee applies to the value of whatever works they design.

In Design–Build, contractors usually seek a predetermined, lump-sum, fixed fee to incorporate within their tenders, and therefore there will be disputes later if the scope of the works and services is not well defined beforehand.

To assist in defining the scope of the engineering works, it is useful to have a check list of works, which may be checked for each project. See, for example, the following list:

Sub-structure
   piling, ground improvements, foundations,
   earthworks, fill,
   retaining walls, water retaining structures,
   ducts, pits, and the like,
   load-bearing walls, non load-bearing masonry walls,
   slabs,
   precast concrete elements.

Super-structure
   external and internal walls, columns, beams, lintels etc., frames, bracing
   floors, slabs, staircases,
   cladding rails, purlins, door posts, wall stiffening, and other supporting
   members,
   roof structures,
   structural aspects of builder's work in connection with subcontractor's
   works,
   and any other works necessary to ensure the continued stability of the
   building and all its parts.

External Works
   earthworks, fill, ground improvements,
   sub-structures, super-structures,
   pits, ducts, tanks,
   roads, paving cross-overs,
   precast elements.

Underground Drainage Systems
   foul, effluent, process, surface water, land drainage, outfalls, pumping
   equipment,
   and including all necessary pipes, culverts, manholes, chambers, soak-
   aways, interceptors, and the like.

Traffic Engineering
   traffic flow rates and patterns on and off site,
   effect on existing infrastructure,
   road and parking layouts.

The engineer and the contractor should examine the project requirements, to determine whether there are any other elements or works to be included.

Tender Stage, Building Services Engineer

Contractors may also make use of a check list for the purpose of agreeing with the building services engineers the services required of them, during the tender stage as follows.

Building Services Engineer's Services, Tender Stage:

  (i)   Examine all available information relating to the M and E engineering works, and discuss with the Contractor and others, the strategy for the preparation of the tender with particular emphasis on the manner in which the M and E engineering information will be presented to enable the Contractor to make an accurate estimate of the cost of the Works.

  (ii) Visit the site to examine and obtain any local information that is reasonably accessible, to assist in the design of the M and E engineering works.

 (iii) Advise on the pre- or post-contract need for the Contractor to obtain or undertake any surveys of existing service installations, either below ground, above ground or within existing buildings to augment the available information to enable the Engineer to carry out his duties during the tender stage, or the post-contract stage.

  (iv) Visit and consult the Local Authority and Statutory Undertakers to ascertain any factors that may influence the M and E engineering design, and the provisions which the Contractor should make in his tender.

  (v) Advise on the scope and nature of the building services (M and E) engineering works required in order to make the Works 'complete' and functional.

  (vi) Advise on the need to employ any other consultants in relation to the M and E engineering works either during the tender period, or post-contract.

 (vii) Assist the Contractor in identifying the need for specialist subcontractors and suggest names of suitable firms; prepare the technical information required in order to seek quotations from such firms; assist in the evaluation of quotations received.

(viii)Provide the Architect with advice and information relating to the M and E engineering works, so that he may coordinate the design development of the Works as a whole during the tender stage.

  (ix) Advise the Contractor on the need for any tests, or the like, which may be required, so that the Contractor may make adequate provision in the tender.

  (x) Advise the Contractor of any conditions or requirements contained within the enquiry which, in the Engineer's opinion, may lead to difficulties, such that it is advisable for the Contractor to limit or qualify his tender.

(xi) Assist the Contractor to prepare a design programme so that realistic allowances may be made for the times for design and approvals.

(xii) Assist the Contractor in assessing reasonable contingencies to be allowed for additional costs which may arise during the post-contract design development.

(xiii) Provide the Contractor with documentation to support his tender, in the form of drawings, specifications and the like, all as agreed with the Contractor.

With the building services, contractors often sub-let the detailed design of the services to a subcontractor. Nevertheless, if the contractor employs the engineer to determine the design parameters and other basic criteria, then the list of services would apply, but the list must be modified by deleting many of the provisions if the contractor only employs the engineer to vet the subcontractor's proposals.

Scope of the Building Services (M & E) Engineering Works

To assist in defining the scope of the M and E engineering works, it is useful to have a standard list of works, which may be checked for each project, for example, as follows:

Mechanical
  incoming supplies,
  boiler and calorifier plant, etc.,
  fuel plant, storage, distribution, etc.,
  heating installations,
  mechanical ventilation systems, all types,
  air conditioning, etc.,
  hot water, steam, condensate systems,
  automatic controls, energy management systems,
  compressed air systems,
  refrigeration, cold stores,
  medical gas services,
  central vacuum cleaning installations,
  refuse collection, disposal systems and installations,
  water treatment, filtration, sterilisation installations,
  food preparation and cooking equipment,
  general plant installations,
  mechanical handling installations,
  window cleaning installations.

Fire Fighting
  automatic sprinklers,
  fire hydrant installations,

    hose reels and dry risers,
    halon protection systems,
    smoke extract systems,
    portable equipment.

Public Health
    drainage above ground, (below ground),
    rainwater disposal systems,
    plumbing installations,
    sanitary appliances,
    sewage pumping,
    effluent treatment and disposal.

Transportation Systems
    conveyors, cranes, hoists,
    escalators,
    passenger and goods lifts,
    scissor lifts,
    vehicle lifts.

Electrical
    incoming services,
    sub-station,
    distribution,
    lighting installations, emergency lighting,
    earth bonding,
    small power installations,
    supplies to mechanical services and plant,
    energy management systems,
    telephone installations,
    public address, recording and call systems,
    listening and viewing systems,
    fire detection and alarm systems,
    surveillance and intruder detection and alarm systems,
    computer systems and information distribution network systems,
    standby power generation and uninterrupted power supply systems;

Tender Stage, Other Consultants' Services

Before, during and after the tender period, the contractor, with his design team, may identify the need to appoint and obtain advice from other consultants. The contractor will want to make provision for the costs of such services in his tender.

Services which are often outside the scope of architects' and engineers' practices, and which are often the subject of a separate appointment include, for example:

(i)   Town Planning advice,
(ii)   legal and financial services,
(iii) geotechnical advice and reports,
(iv)   topographic surveys,
(v)   specialist structural analysis,
(vi)   acoustic surveys and analysis,
(vii) consultancy in connection with curtain walling,
(viii) specialist advice in respect of radioactivity,
(ix)   mechanical handling,
(x)   landscaping,
(xi)   interior decor and furniture design, etc.

The Negotiation Stage

In Chapter 5 we saw that there were a number of different varieties of
Design–Build, and only some had explicit negotiation stages. In practice
however, most projects have a negotiating stage, and thus, the comments
here are almost universally applicable.

After the submission of the tenders, employers often ask contractors to
make changes to their offers, for example, alterations in the spatial arrangements, the addition or omission of work, changes in specification or quality,
changes in the conditions under which the work is to be carried out,
amendments to the programme, and other changes that could have an effect
upon the contractor's price or risk.

It is here that mistakes often occur; with the contractor anxious to secure
the contract, he will agree to amendments, often without realising the full
consequences of a change. To overcome this potential problem, the contractor must desist from making assumptions without full consultation. He
must involve his design team and his buyers, planners etc. as fully as he did
during the tender period.

The simple following example illustrates the point:

*Suppose a contractor's tender proposals included metal cladding on the
outside face of a steel-framed building, and in response to a request from
the employer, the contractor quoted the extra-over price for substituting
full-height brick on one elevation.*

*Suppose then, that the contractor made the adjustment to his price by
applying appropriate rates to the addition of brickwork, and the omission
of the cladding, and submitted his revised price to the employer, without
reference to his structural engineer, assuming that the omission of the
cladding rails would more than compensate for the addition of brick
restraints.*

*The contractor could well have subsequently found that it would be
necessary to incorporate the following changes, for which he had made
no allowance:*

The permissible, or recommended, deflections on the structure are tighter for brickwork than for metal cladding, and so additional cost could be involved in stiffening the structure.

The full cost of the brick restraints could have exceeded the saving in cladding rails, because door openings in the walls could have interrupted the continuity of the restraining rails, where much more substantial door posts would be necessary, again because the brick panels could not accommodate the same deflections as could the metal cladding.

Consultant's Services during the Negotiation Stage

It is important for the contractor to agree, before submitting his tender, the basis on which the consultant's fees, if any, shall be calculated for his services during this stage. The employer will normally want free quotations for the changes without committing himself to the contract with the contractor, and hence the contractor could be at risk for the cost of the fees.

Whilst it is easy to predict that this stage will occur, it is invariably difficult to predetermine the extent of the work involved. If the agreement with the consultant omits any reference to services beyond those necessary for the contractor to prepare his tender, there may be unforeseen, and considerable amounts to pay to the consultant for this work. (Hence, perhaps the occasional reluctance of the contractor to involve the consultants too deeply in these negotiations.)

Preparation of Contract Documents

Invariably, the contract documents are finalised and assembled after the employer has given an order to the contractor to proceed with the detailed design; often to place orders with key subcontractors or suppliers; and not infrequently, to commence the construction work itself.

Hence, and paradoxically, the preparation and assembly of the contract documents may be regarded as a 'post-contract' activity. The topic is covered in Chapter 15, 'Contract Documents'.

14 Tendering

'It's no good lying about the price'

Cervantes [Don Quixote]

KEY TOPICS

- Unorthodox and orthodox enquiries
- Declining to tender
- Budget costing with or without design proposals
- Firm price tendering
- Bid manager
- Bill of quantities
- Dissemination of enquiry details
- Tender strategy
- Programming the tender process
- 'Freezing' the design during the tender
- Meetings and coordination
- Tender documentation

INTRODUCTION

For the contractor, if there is one aspect of Design–Build which stands out alone as being the most radically different to Traditional Contracting, it is the tendering process. We have already seen, in Chapter 11, the extent of the contractor's risks/responsibilities in relation to design, appointment of consultants and subcontractors and pricing. The point at which the contractor offers to accept these responsibilities is when he submits his tender.

The importance of this cannot be stressed too highly; when his tender is submitted, the die is cast; he cannot assume that there are any 'angles' to use subsequently. If the contractor thinks there is, he will invariably 'come unstuck'; there will be no opportunity for the contractor to create 'smokescreens' to overcome deficiencies in his tender, or in his subsequent performance, that may be open to him in other forms of contracting.

Other chapters deal with various aspects related to the tender process; in this chapter we see them brought together to understand how the contractor produces his tender, complete with all the necessary documentation, in the limited time available, with a price that is competitive, yet contains sufficient safeguards against loss. Unfortunately, all too often, the successful contractor is the one whose tender contained the most mistakes, and this is neither good for the contractor, nor in the long run, for the employer.

THE ENQUIRY

Understanding the Nature of the Enquiry

To advertise their capability in Design–Build, contractors often stress their willingness to undertake feasibility studies, propose design solutions, give lump-sum prices and programme, *all without cost or commitment.*

Naturally, potential employers often take advantage of such offers; especially if there is any doubt that the project is to go ahead.

With these doubtful projects the employer often deals with one contractor, and the contractor, knowing that he is in this position, will often oblige the employer. He will do it on the basis that it is worth the speculation if there is a possibility of a negotiated contract to follow. Where he loses out is when he provides the employer with all the information, and then the employer makes use of it, to go out to tender to a number of contractors.

If he is not to be involved in too much speculative work on such schemes the contractor must first develop his intelligence gathering process, and then work out methods of dealing with the doubtful situations.

Gathering Background Information to an Enquiry

If there is any doubt at all in the mind of the contractor as to the seriousness of an enquiry, he should ask the employer for enough information to enable him to decide whether to respond to the enquiry. Genuine employers should never be offended by a contractor who asks for basic information on the background to an enquiry. The contractor is, after all, being asked by the employer to expend quite considerable resources in preparing proposals.

To assist in gathering the right information the contractor may make use of a pre-prepared questionnaire, for example, as follows:

(i)   When was the project first mooted, and by whom?

(ii) How is the project to funded, and is it assured?

(iii) What capital sanction process will there be before the go-ahead can be given?

(iv) Who will make recommendations on the viability of the project, and who will make the ultimate decision?

(v)   Are those decision makers already involved in the project, and does it have their support?

(vi) May the contractor meet the decision makers?

(vii) Has a budget costing for the scheme been established, and by whom? What is the budget, so that the contractor can assess the viability in his own mind; and judge the yardsticks for the design of the proposed project.

(viii) Is a firm price tender, accompanied by drawings and specification required, or will an approximate costing, with or without drawings, suffice at this stage?

(ix) Is the employer seeking proposals from others, if so how many, and what type of organisations are they?

(x) When is the submission required, and in what form? Is there any leeway on the date?

UNORTHODOX ENQUIRIES

Recognising an Unorthodox Enquiry

From the intelligence gathered, a contractor will be able to recognise an unorthodox enquiry. It is one where the request to the contractor for proposals is not well documented; typically the contract conditions will not be known, no tender period will be specified or it will be stupidly short, the brief may be largely verbal and the general feel will be that the employer has not approached the scheme in a well considered and structured way.

Response to an Unorthodox Enquiry

A contractor's decision to respond will normally be based on the following factors:

(i) Likelihood that the project will proceed

(ii) Likelihood of securing the contract

(iii) Level of potential profitability

(iv) Level of potential risk

(v) Resources (all departments) available to deal with the tender

(vi) Resources available to design and construct the works, if successful

(vii) Relationship with the employer or his representative

(viii) Could future opportunities arise from the same source?

The contractor has a number of options in the way he responds to an unorthodox enquiry, including:

(i) Decline to make proposals

(ii) Decline to make proposals, but quote fees for proceeding further

(iii) Give budget costing without design proposals

(iv) Give budget costing with outline design proposals

(v) Submit a full firm price tender and proposals

These alternative responses are considered in the following:

Decline to Make Proposals in Response to an Unorthodox Enquiry

Normally a contractor will not want to spoil his chances of future opportunities by declining to tender, so he must be careful in his approach. He
should avoid the following errors of judgement:

(i) Being off-hand, without giving a good explanation
(ii) Not giving the explanation to the appropriate person
(iii) Leaving a 'junior' to explain the decision to the potential employer
(iv) Leaving it late in the tender period to advise the employer of the
     decision
(v) Decline, without good reason, after expressing very keen interest in
     the project
(vi) Giving the employer the impression that his project is not worthwhile.

Decline to Tender for an Unorthodox Enquiry, but Quote Fees for Design

To be attractive, a contractor's fee proposals may be accompanied by some
design ideas to demonstrate the contractor's abilities, and to whet the
employer's appetite; in addition, the contractor should include a clear
statement as to how he sees the the way the scheme may be progressed.

Usually, such a proposal will show how the processes may be divided
into logical stages, with cut-off points and a breakdown of the fee calculation
for each stage of commitment. At an early stage in the process the contractor
should include provision for a budget costing, and from the employer's point
of view it would be beneficial if that budget were to have a guaranteed
degree of accuracy.

More often than not, a contractor's fee proposal will be judged in
comparison with other firms' proposals, which may be of quite diverse
natures, including perhaps an offer to provide a fully detailed tender and
proposals. So if the fee based proposal is to have any chance of succeeding,
the contractor will have to demonstrate the clear advantages and benefits of
the proposals to the employer.

Budget Costing without Design Proposals (Unorthodox Enquiry)

The least time-consuming way to respond to an unorthodox enquiry is to
give the employer an approximate costing, without preparing design solutions, and often this will be sufficient for an employer's needs at the time.

Design–Build contractors' experienced staff normally have their own
yardstick prices for most types of buildings. For example:

Cost per room for a hotel
m^2 per resident and cost per m^2 for a nursing home
m^2 per person and cost per m^2 for an office
Cost per m^2 for a warehouse or other industrial buildings

Budget with Outline Proposals (Unorthodox Enquiry)

If the contractor has the resources available to augment the budget with outline design proposals he will normally make use of them. With CAD or modern reproduction facilities, attractive proposals can be put together economically, and the contractor's ingenuity may earn him reward. Even though such unorthodox enquiries are often speculative, they also often eventually lead to good projects.

If the contractor is submitting outline drawings he should be more sure of the accuracy of his budget. With the drawings he will be able to produce a 'cost-plan' which will be more accurate than 'square-metre' pricing. If the contractor does not have the resources to do this, most professional quantity surveyors can provide the service quickly and economically.

Such proposals should be accompanied by a statement from the contractor as to how he would propose to take the project to the next and subsequent stages.

Firm Price with Full Specification (Unorthodox Enquiry)

There will be occasions when a contractor makes the decision to submit a full firm price tender with detailed proposals even when the enquiry is unorthodox. He may have discovered that the project will definitely proceed and could very attractive to him. It may be that it is potentially even more attractive than an orthodox enquiry. In such cases the contractor will mobilise a full team and respond in the way that is described in the remainder of this chapter for an orthodox enquiry.

ORTHODOX ENQUIRIES

An orthodox enquiry is one where the format of the enquiry is 'professionally' put together, with a specified tender period, proposed contract terms and a proper brief, albeit that the brief may be concise, as it would be in a contractor-led-design project.

Almost by definition, an employer-led-design enquiry will be orthodox.

These enquiries may be put to one contractor for a negotiated contract, or to more for competitive proposals.

Alternative Responses to an Orthodox Enquiry

In practice, contractors have only two options upon receipt of an orthodox enquiry, namely, to tender, or not to tender; there are no half measures.

Usually a contractor will have had notice of the impending enquiry, and by then he will have normally committed himself to submit a bona fide tender in accordance with the conditions laid down in the enquiry. Certainly, if the employer had followed the procedures described in this book, the contractors would have all been notified and all would have previously agreed to tender; in such circumstances contractors have little option other than to submit a tender. 'Reverse marketing' is a term coined to describe the difficulty the contractor has in explaining to a potential employer that he no longer wishes to tender for the project.

However, if the contractor had not been given notice, nor asked whether he was interested, or if the employer had changed any of the basic conditions which had been previously notified, such as:

an increase in the number of contractors tendering,
a reduction in the tender period,
a significant delay in issuing the tender documents,
changes in the contract terms or conditions,
the introduction of onerous programme requirements or restrictions,

then the contractor would have good reason to decline to tender, if he so wished.

TENDER PREPARATION

Under this heading we consider the production of the tender, including the production of the bill of quantities, but the subject of the choice and appointment of the design consultants is not dealt with in any detail as this was discussed in the preceding chapter.

The Bid Manager

Within the tendering Design–Build contractor's organisation, it is necessary to nominate a 'bid manager'. Whilst many contractors do not recognise this title, the role exists and is understood by most.

Contractors often use the title 'project coordinator', but others use the title 'design manager' to describe the person who fulfils this role. Occasionally a contractor may appoint an estimator to the role, but, in practice, most find that the estimator is too involved in the pricing of the works to be involved in the other aspects, and in addition he often lacks the requisite experience.

The bid manager's role includes the following duties:

(i)   Agree the appointment of consultants
(ii)   Agree to the method of production of the bill of quantities
(iii) Programme the tender production
(iv)   Direct and monitor the design team
(v)   Monitor the production of the bill of quantities, usually on a stage by stage basis
(vi)   Coordinate between the various in-house departments, specialist subcontractors and the consultants
(vii) Seek direction from or give direction to the parties as appropriate

(viii) Liaise with the employer

(ix) Be at the centre of all communication.

Bill of Quantities

Design–Build contractors do not have a magic formula for estimating the cost of works at a level of accuracy required for competitive pricing. It is done through detailed analysis and costing, supported by quotations obtained from subcontractors and suppliers.

A 'take-off' or bill of quantities, in one form or another, is absolutely essential for the pricing of 'measured' direct works and for the quantities needed to obtain quotations from 'domestic' subcontractors and suppliers.

Who Prepares the Bill of Quantities?

At the outset, a contractor must decide his strategy for the take-off; he has the following options:

(i) Appoint a professional quantity surveyor

(ii) Use a specialist in-house department

(iii) Use project quantity surveyors

(iv) Rely upon the estimator

What Form Should the Bill Take?

There is the option to either prepare,

(i) a full SMM bill of quantities,

(ii) a builder's bill of quantities, or

(iii) prepare an elemental cost plan.

There is no one solution to the question as to which is the best way to proceed, each contractor and project is different, but there are a number of points worthy of mention:

(i) An estimator who takes-off his own quantities has a good 'feel' for the job, and he can also save time by making monetary allowances where lengthy descriptions and time consuming take-offs would be necessary if a third party were to produce the bill of quantities.

(ii) On the other hand, if the job is too big for the estimator to retain an understanding of the job as a whole 'in his head', he may lose the feel for the job, and in any parts where he adopts short cuts, he is in danger of incurring omissions.

(iii) Also, if the job is too big or complex, the estimator will not have sufficient time to fulfil both roles.

(iv) Estimators will be able to price a SMM bill more accurately than one which is not prepared to any particular method of measurement, and certainly prices obtained from domestic subcontractors will be more accurate if SMM rules apply.

(v)   However, if a contractor decides that a SMM bill of quantities is what he wants, he may find that unacceptable delays creep in whilst the quantity surveyor is waiting for information of sufficient accuracy to enable him to complete sections of the SMM bill.

(vi) Whatever form the bill takes, its primary function is to ensure that everything that will be required in the works is covered by an item. This usually means that there has to be a section of the bill which 'sweeps up' all outstanding items which have not been measured elsewhere, or are the subject of late information from the designers or specialist subcontractors.

(vii) Preparation of bills of quantities is specialist work, requiring experience and an ability to 'interpret' what may, or will, be required from incomplete details, or rough sketches. This is especially true in Design–Build. Hence contractors are at serious risk if they farm out the bill preparation to project quantity surveyors who have little or no such experience; and experience of being a project quantity surveyor on a Design–Build project definitely does not of itself give a surveyor the requisite experience.

(viii) If the bill is prepared by a number of quantity surveyors on a sectional basis, an appropriately experienced managing quantity surveyor should be appointed to take the overall view and to coordinate the work of the surveyors. He would also be responsible for the 'sweep-up' bill.

(ix) Elemental cost plans are a useful way of measuring and pricing projects, especially when the tender period is short.

However, cost plans often have a good 'feel' about them which can lead to a contractor to give them more credence than they deserve. More often than not, as the design is subsequently developed, the contractor will find many omissions but no over-measured items to compensate. To overcome this, experienced contractors allow both full rates against the items, and in addition, an overall contingency. Contractors should resist the temptation of assuming that just one of these measures will be sufficient in itself to cover any risk.

Appointment of a Professional Quantity Surveyor

If a contractor employs a professional quantity surveyor to prepare the bill of quantities, he must decide upon the terms of the appointment, and the extent of the services that are required.

The contractor may wish to make the appointment on a formal basis, using a set of terms and conditions covering all eventualities, including conditions designed to recompense the contractor should the quantity surveyor's bill of quantities contain any deficiencies, which subsequently cause the contractor loss or expense. This may seem ideal, in theory, but in practice it is rarely achievable, for the following reasons:

(i) A PQS would rarely accept such stringent terms, given the short time available for most tenders, and given the piecemeal way in which the information flows to him.

(ii) A PQS who would accept such terms is likely to charge fees at a level which contractors could be unwilling to pay.

(iii) Even if the appointment were made on this strict basis, with the contractor accepting the consequently high fees, it would inevitably become clear during the course of the tender preparation that the PQS is called upon to interpret from sparse information and sketchy details, and as such he would be entirely justified in disclaiming responsibility for the accuracy of the bill of quantities. This is not to say that he could disclaim responsibility for exercising due skill and care, as would have been the case even with a less formal agreement.

Professional Quantity Surveyor's Professional Indemnity Insurance

It is even arguable that it is unnecessary to ensure that a PQS has profes-sional indemnity insurance to cover claims that a contractor may make against him, because of the unlikelihood of a claim succeeding.

In the light of this, contractors should institute sufficient checks into their procedure to eliminate any gross errors; and to reduce the possibility of any other less substantial errors or omissions from the bill of quantities; remember here that we are considering the quantity surveyor's work, and not the work of the designers from which the quantity surveyor takes-off his bill.

Dissemination of Information

In Traditional Contracting it is almost conceivable that a contractor's estimator could price a tender without reference to any other department within, or outside his company, other than perhaps, his buyers and sub-contractors and suppliers. This is because the rules for Traditional Con-tracting are generally standard, and the only risk that the contractor takes is that he can carry out the work, as described, in the time, and for the rates he quotes. He has no problem with all the Design–Build risks that we saw in Chapter 11. It is therefore much more important in Design–Build than in Traditional Contracting to distribute all relevant information to all relevant in-house departments, and external parties, so that none gives advice in ignorance of any of the requirements of the enquiry.

Dissemination of Information within the Contractor's Offices

Upon receipt of an enquiry, a summary of information can be readily prepared and distributed to the departments within the contractor's organisation, so that they are all aware of the enquiry and the basic details from an early date.

Typically the summary would include:

(i)   Project title, in-house abbreviated title, and tender number
(ii)   Employer and his representatives
(iii)   Date received, and date due in
(iv)   Brief description and approximate value
(v)   Obligatory format for presentation, if any
(vi)   Proposed consultants
(vii) Form of contract and any amendments
(viii) Specified subcontractors, if any
(ix)   Provisional or PC sums, if any
(x)   Responsibility for Statutory Approvals
(xi)   Aspects requiring special attention
(xii) Competition, numbers and names, if known

A standard proforma makes the compilation of such information relatively easy.

Distribution of Enquiry Documents

No matter how lengthy or complex is the enquiry documentation, it is short-sighted to save cost or effort by not distributing it to all those who need it to do their job properly.

Under normal circumstances, the complete set should be distributed externally to:

Architect
Structural engineer
Building services engineers
any other consultant
Quantity surveyor
Any major subcontractors who are part of the team from the outset, on a back-to-back basis
Any major subcontractor who is subsequently brought into the team, on a back-to-back basis

Internally, the contractor will normally distribute the documents, as follows.

Complete set of documents to:
  Bid manager
  Estimator
  Taker-off, if in-house
  Buyer

Partial documents, as necessary to:
Planner
Construction management
Quantity surveyor (not for BOQs, but for commercial appraisal)
Legal advisor
Insurance advisor

The bid manager will define the status of the documents; that is to say he will inform the planner, for example, whether he can proceed with the programme, or whether significant changes are yet to take place which could change the whole character of the tender programme later.

Tender Strategy

The bid manager, no doubt in consultation with his executives and probably the estimator and the design team, will define the tender strategy; it may not be complete at the outset, but it will be developed relatively early during the tender period. The topics that it will cover will include, for example:

(i)   Tender preparation programme
(ii)   Details of the design team and the scope of their services
(ii)   Subcontractors to appoint at the outset on a negotiated back-to-back basis
(iv)   Lines of communication between members of the team, including internal departments and external consultants
(v)   Policy with regard to contact with the employer or his representative
(vi)   Format for the submission documents
(vii) Policy with regard to alternative proposals (design or specification)
(viii) Policy with regard to adhering to the conditions laid down in the enquiry, whether the enquiry says, or not, that tenders may be disqualified if they vary from the conditions laid-down

It is the bid manager's responsibility to ensure that all relevant parties are fully aware of the strategy, and any subsequent changes that may be made to it.

Programming the Tender Process

As all tenders are different, and the design stage reached by employers' design teams may vary considerably from job to job, programming a tender could be the subject of a chapter, or more, in itself. Here we shall deal with it briefly, referring to certain universal principles which apply to almost any tender.

In all tenders the prime end objective, above all else, is to establish a sufficiently accurate tender sum, and with it a description of what is included for that sum. It is worth stressing this point, again and again to any Design–Build tender team.

It is of no use to the team if any of them dwell too long on elements, which have little, or no, material effect on the tender, at the expense of the prime objective.

It follows from this, that satisfying the requirements of the estimator may be considered as one 'path' of the programme in network terms, and the other is the concurrent preparation of the submission documentation. It is usually the pricing path which is the critical one, and so we shall consider that in detail.

Programme for Pricing the Works

The pricing of the works may be broken down into five sections for consideration, namely:

(i) Work to be designed and priced by specialist subcontractors, e.g.,
piling and ground improvement,
structural steel or precast concrete frames,
curtain walling,
building services.

(ii) Items to be measured, then priced by obtaining quotations from subcontractors or suppliers, e.g.,
disposal of excavated materials,
formwork and steelfixing,
roof coverings,
brick and blockwork,
plastering and screeding,
carpentry and joinery,
floor wall and ceiling finishes,
decoration,
various external works, pavings, etc.

(iii) Work to be quantified and priced by subcontractors on the basis of drawings and a specifications supplied to them by the contractor, where little or no design is required of them, other than the design which is inherent in their 'product', e.g.,
windows and simple curtain walling,
roof sheeting and side cladding,
precast floors and systems,
lifts, hoists and escalators,
suspended ceilings,
furniture and fittings.

(iv) Items to be measured then priced, in-house, by the estimator, e.g.,
excavation,
concreting.

(v) Preliminaries, fees, etc.

Broadly speaking, these categories have been listed in descending order of the time that a contractor will take, from the receipt of the bill of quantities or drawings and specifications, to price, or obtain prices for the work. It is not possible to suggest the time that each will take, because that will vary upon the characteristics of each job, but it is possible to generalise with a notional programme showing the relationships between each category, in a not-to-scale diagram, see Figure 14.1.

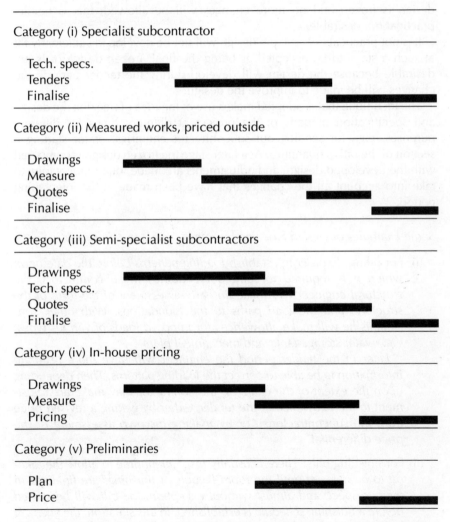

Figure 14.1 Bar chart showing the notional relationship between the five categories of work for pricing

Using the philosophy shown in Figure 14.1 a contractor can separate the trades into the appropriate categories, and programme each according to its characteristics.

Allowances must be made at the start of the programme for briefing, and at the end for the finalisation and settlement of the tender, as a whole.

The Concept of 'Freezing' the Design

In preparing Design–Build tenders great use may be made of the concept of freezing the design. This is not simply to say at a particular stage that the design is frozen and no more changes will be allowed; this is neither practical nor desirable.

It is not practical because the design will not be developed sufficiently far at such a stage to be accepted as being the final, frozen design; it is not desirable because the design will develop during the tender period, and changes will be made to improve the design.

The technique that is adopted makes use of the idea of freezing the design and specification of parts of the project at different times for different purposes, e.g., for sending out subcontractor enquiries, or for starting a section of the bill of quantities. At a later stage the frozen design is compared with the developed design, and adjustments are made, often monetarily, to take into account all the changes that have been made in the intervening period.

Some Examples of Design Freezing by Stages, and Late Adjustment

(i) For piling: *In order to establish loading patterns before the date upon which it is required to send out enquiries to subcontractors, the structural engineer will make an early assessment of the form of the structure, and the load paths to the foundations. With other consultants he will make allowances for imposed loads of various types; e.g., water storage tanks and mechanical plant.*

Later in the tender period the engineer will have more accurate information to be able to correct the loading patterns. Then depending upon the extent of the changes, the contractor may make an adjustment to the allowance in the tender, either by getting a revised price from the subcontractor(s), or by making his own assessment of the price differential.

(ii) For finishing bills: *There is usually insufficient time to allow the take-off to be delayed until the room layouts or finishings are finalised in every respect, and almost without exception, the bill will be started before a finishing schedule is established. In this situation, the taker-off will use those drawings and other information which is available at the time and mark them up to show what assumptions he has made, either in consultation with others in the team or not; it is better to make an assumption and forge ahead, than to await a decision and get the bills out late. The estimator can then use the bills to seek quotations as necessary; and as the specifications for finishings are finalised (and changed?), and the subcontract quotes come in, the estimator can make the necessary adjustments.*

To adopt this sectional freezing approach demands a high degree of control, and it is in the management of this process, perhaps, that the bid manager's skills are most needed and tested.

MEETINGS AND COORDINATION DURING THE TENDER PERIOD

The bid manager will normally convene, chair and be responsible for the minutes of all tender progress and coordination meetings. The management of this is an art that comes with experience; it is difficult to strike the happy medium between too many meetings with too many attendees on the one hand, and too few meetings with inadequate representation from all the interested parties on the other hand. The former are cumbersome and time consuming and rarely produce positive results, whilst with the latter there is a danger of leaving people to work in isolation, ignorant of the requirements or the decisions of others; short-handed meetings can also waste the time of those present when advice is needed from an absent party, and without it, progress cannot be made.

Minutes of Meetings During the Tender Period

The purpose of meetings in general is to record decisions, agreements or facts, so that all parties have the same understanding as to how they should proceed with their duties. For this reason they are a very important management tool in managing the tender process.

However, the bid manager or his delegate could spend too much of their time doing little other than writing minutes, and by the time they are distributed and read by the parties the information contained in them could be out of date, and not only that, if they are lengthy the parties will have little time to read them. In the event of a problem arising it is too late afterwards and serves no purpose for the bid manager to point to the minutes to identify the culprit; in real terms what recompense can he obtain if the error led to the loss of the project for example?

There are a number of different meetings required during the tender period and these are described in the following part of this chapter.

Internal Briefing Meeting

The purpose of this meeting is to bring together all in-house members of the bid team, specifically to:

(i)   Discuss the enquiry documentation (previously circulated)
(ii) Agree/confirm the design team, if not already done
(iii) Outline to the meeting, the tender preparation programme and seek comments from all parties
(iv) Agree tender procedures, and 'who does what'
(v)   Raise and discuss any initial queries
(vi) Discuss the question of specialist subcontractors
(vii) Preliminary discussion on the submission format
(viii)Discuss any particularly features of the enquiry which may cause difficulties of any sort
(ix) Agree any cost restraints for the tender preparation.

External Briefing Meeting

The purpose of this meeting is to bring together the designers, the taker-off and key members of the contractor's organisation and:

(i)   Distribute enquiry information, if not already done
(ii) For the bid manager to describe the contractor's bid strategy, and to invite comment upon it
(iii) For each member of the design team to introduce key personnel and to explain his organisation's modus operandi
(iv) Discuss the enquiry, and any background information that will assist in the interpretation of the employer's requirements
(v)   Discuss the tender preparation programme, and obtain the agreement of all parties to it
(vi) Preliminary discussions on specialist subcontractors
(vii) Agree timetable for coordination meetings
(viii)Discuss presentation format.

Tender Progress and Design Coordination Meetings

These meetings are held regularly and are chaired by the bid manager. The meetings are not the place for resolving detailed design questions. They deal with more strategic issues. They seek, for example, to ensure that proper coordination between the design team is taking place, but without being the actual vehicle through which the coordination takes place.

Specifically the purpose is to :

(i)   Monitor the progress of the design team
(ii) Monitor the progress of the technical specifications for subcontract enquiries
(iii) Monitor the progress of the take-off
(iv) Discuss design development and options
(v)   Check the effectiveness of the coordination between disciplines
(vi) Discuss strategic and tactical questions
(vii) Monitor the preparation of the proposal documents.

Whilst it is difficult to generalise on an agenda, this list could well provide the framework for controlling the direction of the discussions.

Commonly, the minutes of the previous meeting form the basis for discussion, with all parties being invited to raise new issues, and such subjects are then incorporated into the subsequent meetings via the minutes.

To do justice to all the key points requires firm control on the part of the bid manager, as it is common for discussions to become repetitive, over-long and too detailed.

Venue of the Tender Progress and Coordination Meetings

It is essential that the meetings are held where communications are readily to hand (telephone and fax), so that contact may be made from time to time, during the meeting if necessary, to others involved in the tender. It is also vital that there are copying facilities available to assist in the issue to the parties, of information prepared at the meeting.

To have these facilities available means that the venue should normally be at the office of one of the parties involved.

Detailed Design Coordination Meetings

These meetings will be held on an adhoc basis, involving just those members involved in the specific topics under discussion.

Minutes of these meetings normally take the form of an abridged list of decisions made, and questions yet to be answered. They could be distributed to all parties by fax, or at the progress meeting if that is early enough.

'Sweep-up' Meeting

This meeting is held at the end of the tender period, but before the contractor's finalisation meeting.

To be effective, it must be attended by representatives of the complete design team, the taker-off and the estimator. It will normally be chaired by the bid manager.

The representatives must be those who were actually involved in the tender, not figureheads, so that when questioned as to what may have been assumed in a particular decision, the answer 'I assumed ...', or 'I allowed for...' can be given, and not 'I am not sure, I will ask my junior who worked on that part of the works'.

The purpose of the meeting is threefold, namely:

(i) To seek out errors and omissions, and correct them
(ii) To seek out over-provision or duplication
(iii) To assess the risk of additional costs arising in the post-contract stage.

Format for a Sweep-up Meeting

At the meeting, the bid manager will introduce each element of the bill of quantities in turn and ask, for *each element,* or group of items:

(i)   The estimator to describe briefly what he has priced
(ii) The taker-off what assumptions he has made, and in principle what his source of information was
(iii) The architect and engineers are asked whether the answers 'sound right' to them, and whether they have any last thoughts, which could improve the tender, or which may affect the risk
(iv) The architect, the engineers, the taker-off and the estimator are asked to discuss what subsequent problems or extras could arise; and between them to quantify them in monetary terms.

    For example, *if the proposed contract terms lay the risk of unforeseen ground conditions with the contractor, then the meeting would consider the weight of the risk by assessing possible volumes of soft spots for example, and apply a rate to the potential excavation, cart-away and imported fill.*

Representatives should be encouraged to cast doubt on their own work, so that errors will be found before it is too late. Defensive stances must not be tolerated.

The estimator will keep the tally of the omissions and additions.

The monetary values for the risk assessment would be tallied, but kept separate, because they should not be treated as straight additions.

In addition to the individually assessed risks, the team should discuss the overall risk, taking into account the nature of the project, the general 'feel' for the way in which the tender has been developed, the contract terms, etc. and from that the team should discuss the contingency that they would recommend for both the individual elements, and for the overall situation.

The bid manager, or estimator will report on the findings at the contractor's finalisation meeting, so that the executive may make a valued judgement upon the contingency to be included. This contingency must not be confused with the general contracting risks which are common to the contractor in both Traditional Contracting and Design–Build.

TENDER DOCUMENTATION

Employer-led-Design Projects

In an employer-led-design project, the format for the tender submission will invariably be laid down in the enquiry, and contractors generally adhere to these instructions.

A contractor may consider that this is a minimum, and that he may have a better chance of success if he provides further information to enable his proposals to be evaluated more fully. He will certainly wish to highlight any aspects of his proposals which he feels are innovative, or which would provide the employer with a better standard than was specified in the enquiry.

Contractor-led-Design Projects

Most contractor-led-design tender proposals are bound documents, normally attractively presented. Drawings are sometimes in A3 format, bound within a brochure, or otherwise of any suitable size and folded, or both.

Typically, the contractor's tender, or proposals, will contain the following information:

Introductory

Title page, with date and reference
Contents page
Introduction, explaining the background to the enquiry

Contractual

Contract terms, conditions, and details
Basis of proposals, brief overall description, scope of work, floor areas, etc.

Proposal Drawings

Architectural	Plans of all floors and roof	1:100
	All elevations	1:100, 1:50
	Principal sections	1:50
	Some detailed plans, e.g. toilets	1:20
	Details of particular features	1:20, 1:5
	External works	1:200,1:100
Structural	Foundation, general arrangement	1:100
	Structure, general arrangement	1:100
	External works construction	Various
Mechanical	Radiator/heater layouts	1:100
services	Main pipe routes	1:100
	H & C water supply routes	1:100
	Tank sizes and locations	1:100
	Boiler/calorifier layout	NTS
	A/C and vent plant layout	NTS
	Duct runs	1:100
	Schematics, all systems	NTS
Electrical	Distribution diagrams	NTS
services	Floor layouts showing:	1:100
	lighting and emergency lighting	
	small power	
	fire alarms	
	public address, telephone, TV, etc.	
	distribution boards	

Proposal Specifications

Particular specifications for all trades, by description, product name or by performance

General specification for workmanship and materials for all trades

Other Information

Programme and method statement

Appendices

Further information, for example, the contractor's organisation and experience of similar works

The Offer

Contractors normally make their offers in a separate letter or bound document containing all the financial details.

15 Contract Documents – Contractor's Action

'And where, though all things differ,
all agree'

A Pope [*Windsor Forest*]

KEY TOPICS

- A contract = offer and acceptance
- Letters of intent
- Preparation of JCT 1981 Contract Documents (contractor's involvement)

  Employer's Requirements
  Contractor's Proposals
  Contract Sum Analysis
  Optional clauses
  Appendices

- Preparation of BPF/ACA Contract Documents (contractor's involvement)

  Contract Drawings
  Specification
  Time Schedule
  Schedule of Activities
  (Contract Bills)

INTRODUCTION

We have seen in Chapter 4 'Forms of Contract' that the author of the various parts of the contract documentation varies from one form of contract to another; in this chapter we concentrate on the way in which contractors deal with, and check, contract documentation.

The most common form of Design–Build contract is the JCT 1981 Form and this will be dealt with most comprehensively. Reference to the BPF Form follows at the end of the chapter.

A Contract Consists of an Offer and Acceptance

It is useful to remember that no contract exists until there is an 'offer' and 'acceptance' of that offer. Generally in Design–Build contracting, a contractor makes an offer which the employer will not accept immediately.

Almost without exception, negotiations take place after a contractor submits his tender, up to the point when the employer and contractor have an agreed understanding of what the employer requires, how the contractor proposes to meet those requirements, and what the payment terms and contractual conditions will be.

Because of this, the employer's requirements, as contained in the *enquiry*, and the contractor's proposals, as contained in the *tender*, do not normally constitute the 'offer and acceptance'.

Relative Importance of Design–Build Contract Documents

It is probably true to say that the Design–Build contract documents are generally more comprehensive than their Traditional Contracting counterpart.

This is because:

(i) In Traditional Contracting, the contract documents describe the general nature of the work and, by way of the bill of quantities, describe the quantity and nature of the work that the consultants envisage will be required. *The actual work that the contractor undertakes is that which is shown on the working drawings, which may, or may not be the same as is shown on the contract drawings.* Hence, surprising as it sounds, the contract drawings can often be completely ignored by the parties in a Traditional Contract.

(ii) With Design–Build the contract documents define precisely what it is that the contractor is to build, and therefore errors and discrepancies within Design–Build contract documents can have a more serious effect on the employer, the contractor, and even on the consultants than any such problem encountered in Traditional Contract documents.

LETTERS OF INTENT

Once an employer and a contractor have reached agreement in principle, the employer will normally issue a letter, to the contractor, instructing him to proceed with design on an agreed but limited basis. This letter is usually called a 'letter of intent'.

At the same time both parties will be preparing their own parts of the contract documents.

Before acting upon the letter of intent, the contractor should satisfy himself that his costs and commitments are covered by the letter, and that the letter is not merely an expression of intent by the employer which says, in effect no more than, 'all being well' he will enter into contract with the contractor in due course.

The more specific the letter is, the better it is for all concerned.

The contractor should check the meaning of the letter carefully to satisfy himself on the following questions:

(i) Upon receipt of a letter of intent for preparatory design works, the contractor will need to give his consultants instructions to proceed. Are the terms of the letter of intent such that the contractor may enter into a back-to-back agreement with his consultants and the sub-contractors whose design must be progressed?

(ii) Does the letter of intent cover for ordering materials?

(ii) What happens if the employer calls a halt to the work before completion of the work authorised by the letter of intent? Is there a mechanism for covering partial costs?

(iv) What happens if the authorisation covers a specific amount of work, to be completed by a specific date, and the contractor fails to complete that work within the allotted time, and the employer calls a halt prior to completion of the work? What payment rules would apply in this eventuality?

(v) Furthermore, what if the employer does not call a halt to the work but the delay has a knock-on effect on the dates achievable for the main contract and the contractor's estimated costs? Will the employer or the contractor accept the penalties?

CONTRACT DOCUMENTS – JCT 1981 FORM OF CONTRACT

The documents which are referred to in the contract, and therefore those which must be prepared, are as follows:

Employer prepares	*Contractor prepares*
Employer's requirements	Contractor's Proposals
Contract clause options	Contract Sum Analysis
Appendix (part)	Appendix (part)

Documents Prepared by the Employer

Employer's Requirements

Often, in a contractor-led-design project, the employer will have not prepared a brief which indicates his requirements in a way which could be capable of being used as the Employer's Requirements, and in such cases they may be prepared, jointly, by the contractor and the employer.

Otherwise, it is much more normal for the Employer's Requirements to be prepared by the employer. It (note the singular, because it is the name of a contract document, as opposed to a set of requirements) will usually be based on the documents which the employer prepared for the enquiry, whether or not the enquiry was for an employer-led-, or contractor-led-design project.

When presented with the Employer's Requirements, the contractor should check them for:

(i) discrepancies within the document itself,
(ii) discrepancies between it and the Contractor's Proposals,
(ii) elements or requirements which may have been introduced and not provided for in the Contract Sum,
(iv) the inclusion of any provisional sums which have been agreed with the Employer as the basis for calculating payment for such items.

Needless to say, once a contract is signed, the contractor will have no redress if he failed to notice anything contained in the Employer's Requirements that could cause him loss and expense for which he had made no provision in the contract sum.

Contract Clauses – Employer's Options, (JCT 81)

In the enquiry document, or in the contractor's offer, the terms of the contract should be stated, and it will have been upon this basis that the contractor assessed his risk and his tender, and hence the contract sum. The contractor must therefore check that the proposed contract, presented to him by the employer for signature, contains no changes in the terms and conditions that may involve him in additional risk or cost.

Appendices (Employer's proposals)

It is not uncommon for the appendices to a contract to be incomplete until a quite late stage in negotiations. Clearly this can have a marked affect on a contractor's risks. A contractor should endeavour to establish at the earliest possible stage what the employer proposes, or will accept, because if it is left too late, the contractor can well lose bargaining power. It must be added though, that it is sometimes to the contractor's advantage, because in some situations it is the employer who loses the initiative in the negotiations by leaving these questions too late.

Notwithstanding this, contractors should check that the contents of the appendices, as presented, are consistent with any previous details given.

Documents Prepared by the Contractor (JCT 81)

The Contractor's Proposals

The Contractor's Proposals, a contract document, normally consist of four parts, namely:

(i) Drawings
(ii) Schedules
(ii) Particular Specification and Scope of Works
(iv) General Specification (Optional).

Contractor's Proposals – Drawings

To a large extent, the drawings and specifications contained within the Contractor's Proposals will be dictated by the contents of the Employer's Requirements, However, the following lists would be typical for most projects, unless the employer's drawings and specifications are in sufficient detail and not superseded by proposals of the contractor, in which case it would be unnecessary to duplicate the details in the Contractor's Proposals.

Typical Contractor's Proposals – Drawings (JCT 1981)

Architectural Drawings

Floor plans	1:100
Roof plan	1:100
Detail plans	1:50 & 1:20
Sections	1:50
Detail sections	1:20
Elevations	1:50
Elevation details	1:20
Overall Site plan	1:100 or 1:200 or 1:500
Detail site plans	1:100 or 1:200
Landscaping – soft	as required
Details external elements	various scales

Depending upon the type of building, there may be the need for plans, sections, elevations and details of internal fittings and equipment.

There may also be a need for drawings to indicate the treatment of such things as the concealment of services.

Types of floor, wall and ceiling finishes and joinery will often not be illustrated on drawings, but be included in schedules, as shown in the following section.

Structural Drawings

It is normal to include sufficient structural drawings in the Contractor's Proposals to indicate the nature and form of the structure and substructure, without normally including any RC details or steelwork fabrication details. Typically the list of drawings will be as follows:

Piling layout (if any)	1:100
Sub-structure general arrangements	1:100
Framing plans	1:100
Floor plans	1:100
Roof plan	1:100
Sections	various scales
External structures	various scales
Hard surfacing details	various scales
Underground drainage	1:100–1:200
Drainage details	various scales
Outfall details	various scales

Mechanical Installations Drawings

Even at this stage, many of the details of the heating, ventilation, air-conditioning, hot and cold water, and sanitary systems will not be resolved, and much reliance will be placed on schematic drawings and the particular specification for the definition of the majority of the systems.

Notwithstanding this, typical Contractor's Proposals contain the following drawings:

Schematic diagrams of all systems.		Not to scale
Plant rooms schematic layout		1:20–1:50
Floor plans	– pipe runs	1:100
	– heater positions	1:100
	– duct runs	1:100
	– grille positions	1:100
	– main risers	1:100
	– soil and vent pipes	1:100
Details of heaters, etc.		as required

Electrical Drawings

Distribution diagrams of all systems		Not to scale
High voltage distribution		Various scales
Medium voltage distribution		Various scales
Layouts	– lighting	1:100
	– emergency lighting	1:100
	– small power	1:100
	– communications	1:100
	– low voltage	1:100
	– alarm systems	1:100
	– security systems	1:100
Details of fittings		as required

Typical Contractor's Proposals – Schedules (JCT 81)

Schedules are a convenient way of describing many of the elements of a project, and Contractor's Proposals may typically contain the following:

Room schedule (indicating size, use, location, etc.)
Schedules describing for each room or area:
    floor finishes
    wall finishes
    ceiling finishes
    fixtures and fittings.

The following schedules may be included, or otherwise dealt with in a more general way in the particular specification:

Door schedule
Ironmongery schedule
Window schedule.

Various elements of the servicing of the building may be described by way of schedules for example:

Room design temperatures
Air change rates
Lighting levels
Equipment schedules
Schedules of supplies to employer's equipment.

Particular Specification and Scope of Work – Contractor's Proposals

The particular specification should describe the works fully and describe all that the contractor intends to incorporate within the works. It should refer to the drawings and schedules, as appropriate, and also to the general specification.

The contractor must ensure that there are no discrepancies within this document or between this document and any other contract document.

The conditions of the contract will determine the consequences of any discrepancy, and who meets extra costs, should they arise; even if the employer becomes liable, it is far more satisfactory if time and trouble is taken by the contractor, to check the document diligently to avoid the problem in the first place.

It is in this particular document that the contractor has his last opportunity to lay down the specification of the building and all its constituent parts, and generally it is the contractor who decides the precise amount of detail which is included. From time to time one hears the view expressed that, from a contractor's point of view, it is better to keep the specification as loose and general as possible, because this will give him scope during the course of the contract to effect savings by choosing less expensive options than he had provided for in the contract sum.

This is a risky approach which will often back-fire. It is the employer who 'holds the purse string' and so he has the power which is difficult to challenge. He can deduct amounts, from payments due, if he is not satisfied with what is being provided in circumstances where the contract documents are vague. This would be especially true if the Employer's Requirements contains a clause of the type which states that the contractor 'will provide materials, etc. of the best standard available' or 'that will be fit for the purpose intended'.

The particular specification and scope of works contained within the Contractor's Proposals will invariably be based upon the tender proposals; amplified as appropriate and amended to accord with any post-tender agreements.

Contract Sum Analysis – Contractor's Proposals (JCT 81)

Usually a breakdown of the tender figure is included within the contractor's tender proposals, or sometimes this may be submitted during post-tender negotiations. The purpose of the breakdown is for the employer to make comparisons between tenders on a like-for-like basis. It also enables the employer and the contractor to look at the figures together to seek savings in cost.

The Contract Sum Analysis, a contract document, has different purposes; in a practical sense it may likened to a bill of quantities, in that under the terms of the Contract it is to be used for:

> valuing variations, and
> calculating fluctuations (if fluctuations apply).

In practice, it is also used as a basis for calculating monthly valuations, although the Contract does not stipulate that it should be.

CONTRACT DOCUMENTS – BPF/ACA CONTRACT

Preamble C of the BPF/ACA Form of Building Agreement 1984 specifies the contract documents, as follows:

(i) Certain drawings
(ii) Time Schedule

Optional

(iii) Schedule of Activities
(iv) A specification
(v) Bills of quantities

The option of including a bill of quantities is rarely used.

BPF/ACA Contract Drawings and Specification

It is the intention of the BPF System that the contract drawings and specification are prepared by the client or his consultants. There is no provision for 'Contractor's Proposals' (cf. the JCT 1981 Form), although in practice, a contractor may be appointed by way of a letter of intent in the pre-contract period and, by agreement, prepare some of the drawings and specification for the contract documents.

In whatever way the drawings and specification are prepared they should fully describe the works, or lay down the parameters within which the contractor shall design the remaining parts of the works.

Upon receipt of proposed contract drawings and specifications prepared by the client, the contractor should ensure that he has everything covered in the contract sum.

If a contractor assists in the preparation of the contract drawings and specification, then, obviously, he should ensure that they contain nothing for which he has no allowance in the contract sum. If the contractor prepares the contract drawings and specification, he should note that he may well have a greater obligation to the client, and hence risk, than is generally envisaged by the terms of the BPF/ACA Form.

BPF/ACA Time Schedule

The Time Schedule is not a programme, as its name may imply. It is in two parts namely:

(i) date(s) for possession and completion,
(ii) schedule of drawings and details etc. which the contractor is to provide to the Client's Representative (for sanction) giving dates for each.

Clients normally specify the dates for possession and completion, and contractors should check that they are in accordance with what they had envisaged.

The contractor produces the Schedule, with dates for submission to the client, of drawings and details, and the period for return by the client, if other than 10 working days.

There is no absolute obligation upon contractors to ensure that the list is comprehensive in so far as the Contract (Clause 2) allows, or requires, the Contractor to prepare and submit 'other drawings', etc. to the Client's Representative.

BPF/ACA Schedule of Activities

The BPF Manual gives guidance on the make-up of the Schedule of Activities, suggesting that, for each activity, the following information should be given:

(i) Activity description
(ii) Quantity
(iii) Resources
(iv) Start time (week number)
(v) Duration
(vi) Price.

The Schedule is to be prepared by the contractor, often in an abbreviated form as part of his tender.

The BPF Manual gives typical examples of the Schedule, in an abbreviated form to be submitted with the tender, and in a developed form for use during the contract. It distinguishes between the treatment of 'measured', or measurable, work and 'non-measured' elements such as preliminary costs.

The Contract provides for interim payments to be based only upon *completed* activities as defined and described in the Schedule. Thus it is in contractors' interests to pre-plan the works well but also to break down the Schedule into items which are as small as practical.

BPF/ACA Contract Bills

The BPF System is so drafted that a bill of quantities can be used for the tendering and valuation of part, or all, of the works. Having said this, it is the stated intention of the authors of the BPF System that bills should not normally be used, and that the lump-sum concept should apply.

16 Contractor's Consultants' Post-Contract Services

'Entire and whole
and perfect the service...'

Sir Cecil Spring-Rice [*I vow to thee my Country*]

KEY TOPICS

- Architect's, structural and building services engineers' services:

  Design
  Design coordination
  Supervision and inspection
  Administration
     Contractor's design management
     Contractor's consultants' fees

INTRODUCTION

This chapter considers the services which contractors require from their consultants in the post-contract period; that is, during the lead-in, construction and maintenance periods. It does not cover the terms of the agreements. These have already been discussed in Chapter 12, 'Selection and Appointment of the Contractor's Consultants'.

Whereas this chapter is directed particularly at those contractors who employ independent consultants, much of the contents could apply to a contractor with in-house designers.

Many contractors will have their own standard list of services to be provided by consultants and they will have been tailored to complement the contribution which the contractor himself wishes to make.

Some contractors may engage consultants merely as providers of a drafting service; they do this to keep costs to a minimum. Others regard this as a dangerous and short-sighted practice and employ consultants to give a more comprehensive service. It is this latter practice which this chapter envisages.

Design–Build contractor's consultants' post-contract services may be divided into four categories for the purpose of this chapter, as follows:

(i) Design
(ii) Design coordination

(iii) Supervision and inspection
(iv) Administration.

They are considered in turn for architects, structural engineers and building services consultants.

CONTRACTOR'S ARCHITECT'S POST-CONTRACT SERVICES

The RIBA Architect's Appointment divides architects' services into three parts, namely:

(i) Preliminary Services
(ii) Basic Services
(ii) Other Services.

The basic services are further subdivided into Stages A to L, and these are formulated to suit the sequence of operations of Traditional Contracting. In Design–Build, the stages vary in both timing and content.

In Chapter 13, we saw the services required by contractors in the pre-contract stage, here we consider the post-contract stage.

Contractor's Architect's – Post-Contract Design

On most projects contractors require their architect to:

(i)   Make amendments as necessary to drawings so that they can be incorporated into the Contract Documents
(ii) Re-examine the project requirements to establish the Employer's brief and the Contractor's obligations. (This may seem to be a duplication of previous work, but in practice it is an important step which represents a formal commencement of the post-contract design process)
(iii) Develop the detailed design accordingly
(iv) Assist in monitoring the design development against the tender or Contract provisions
(v)   Consult the Local and other Statutory Authorities, and make application for approvals as required by the Contract
(vi) Participate in any sanctioning process that is required by the Contract
(vii) Provide the Contractor with performance specifications or other information, to enable the Contractor to obtain quotations from specialist subcontractors
(viii) Check specialist subcontractor's proposals to ensure that they are technically suitable
(ix) Prepare working drawings, details, schedules, and other information that the Contractor will need in order to construct the Works
(x)   Assist in the preparation of maintenance manuals
(xi) Prepare as-built drawings.

Contractor's Architect's – Post-Contract Design Coordination

Usually contractors appoint the architect to be the design team leader, as this is the role that he is trained to do and he fulfils the role in Traditional Contracting. A 'builder' is not qualified for this role as he would normally have inadequate design experience. If the architect is appointed thus, his duties will include the following:

(i) Assist in the preparation of the design programme showing both the dates for issue of information to the Contractor, and also the dates upon which information will be required by the consultants in order to undertake their own design work.
(ii) Coordinate the design work undertaken by all others including both consultants and specialist subcontractors. This involves checking that all designers are working to the same parameters, and that spatially all elements 'fit'.
(iii) Coordinate the production of as-built drawing by other designers, into a comprehensive and compatible set of documents.

Contractor's Architect's – Supervision and Inspection

In order to ensure that the architect takes full responsibility for his design work, it is safer to require (and not prevent) him to inspect the works as they are constructed. Without this provision he may justifiably avoid liability for defects on the grounds that the intent of his design was not carried through into the construction and without inspections he had no opportunity to prevent it. Hence many contractors employ their architects to:

(i) Inspect the construction of the Works on a visiting or full-time resident basis, depending upon the nature of the work
(ii) Visit manufacturers' premises to check on quality
(iii) Witness tests
(iv) Make inspections upon Practical Completion and at the end of the Maintenance Period.

Contractor's Architect's – Post-Contract Administration

Contractors may sub-let to their architect some of the administration work involved in the management of the project, for example:

(i) Minutes of meetings
(ii) Chairing various meetings
(iii) Prepararation of various reports, progress, for example.

CONTRACTOR'S STRUCTURAL ENGINEER'S POST-CONTRACT SERVICES

Normally an engineer's appointment would have been defined at the tender stage, and so the post-contract appointment would be a continuation of that agreement, but occasionally, the engineer is not engaged until the contractor has secured the contract and so the services described in this chapter have been formulated to cover both situations.

The ACE (Association of Consulting Engineers) Conditions of Engagement (Agreement 3, 1984), cover the terms of appointment and services to be provided by a structural engineer for a project where an architect is appointed by the client and is harmonised with the RIBA Architect's Appointment. As such, the ACE conditions are not entirely suited to the Design–Build situation, and therefore contractors must reconsider the services which they require for any particular project.

Scope of the Engineering Works

To define the services, it is first necessary to define the actual scope of work for which the engineer is to provide his service. A check list of the scope of the engineering works is included in Chapter 13, and if the scope has not been previously settled, that check list may be used at this stage for the purpose.

Contractor's Structural Engineers – Post-Contract Service

It is not convenient to subdivide structural engineers' work into categories; design, coordination, etc., as it was with the architect. In the engineer's case the scope of the post-contract services will be considered as a whole.

On most projects the contractor will, in respect of the works defined in the 'scope of the works', require the structural engineer to carry out the following work:

Paragraphs in italics may not be required if the work was undertaken satisfactorily during the pre-contract stages.

(i)   Prepare, or amend, any drawings or details needed for incorporation into the Contract Documents

(ii) *Examine with the Contractor and other members of the design team, all available information in order to establish the structural engineering requirements of the Works*

(iii) Assist in the preparation of the design and information release programme

(iv) *Visit the site and make enquiries to determine any factors which could affect the engineering design*

(v)   *Advise on the need for topographic or soil surveys*

(vi)   *Consult the Local or any other relevant Authority on matters which may affect the design of the engineering work*

(vii)   *Advise on the need for any other specialist consultant's advice in connection with the engineering works*

(viii) *Liaise with the Contractor to determine the scope of the engineering works to be sublet to specialist subcontractors, and provide all information which will be required to enable tenders to be obtained from such subcontractors. Assist in the evaluation of tenders received*

(ix)   Provide the Architect with information which he needs to co-ordinate the design of the works

(x)   *Advise on the need for any tests that may be required,* provide the necessary instructions, and as necessary witness and report upon such tests

(xi)   Provide information and assist as necessary in obtaining approvals and sanctions required by the terms of the Contract

(xii)   Provide the Architect with any engineering information required in relation to builder's work in connection with specialist sub-contractors' works

(xiii) Examine specialist subcontractors' proposals and working drawings for consistency with the intent of the engineering design and specification

(xiv)   Prepare working drawings, details, specifications and schedules which the Contractor will require to construct the engineering works

(xv)   Inspect the Works either by visiting, or by employing a resident engineer

(xvi)   Attend meetings as required by the Contractor

(xvii) Assist in practical and final completion inspections of specialist sub-contractors' engineering works

(xviii) Assist the contractor by providing technical advice in the event of any dispute arising, either with the Employer, or with any specialist engineering subcontractor.

CONTRACTOR'S BUILDING SERVICES ENGINEERS' POST-CONTRACT CONSULTANCY

Building services engineers may be employed contractors to either:

(i)   prepare performance specifications against which specialist sub-contractors quote for the detailed design and installation of the work, or

(ii)   undertake the detailed design of the work for the subcontractor to install only.

As a further alternative, contractors may require building services engineers to design some or parts of the works leaving the subcontractor to complete the design of the remainder.

The consultancy service required by a contractor depends upon the approach he decides to adopt. In either case the scope of the engineering works is to be defined, and that will normally be similar to the scope agreed at the pre-tender stage. A check list of the scope of building services works is contained in Chapter 13, 'Contractor's Consultants' Pre-contract Services'.

Contractor's Building Services Engineer – Post-Contract Consultancy

Subcontractor-designed Systems

If subcontractors are to be responsible for the design of the installations, the building services consultant, if appointed at all, will have only a supervisory and watching brief which may include the following:

(i)   Assist in compiling the Contract Drawings and Specifications

(ii) Examine the subcontractor's drawings and specifications as they are prepared during the design development stage, and comment as appropriate in respect of
compliance with the Contract requirements and intent
compliance with regulations
compliance with good practice
'grey areas' and completeness

(iii) Visit the site (or provide a resident inspector) to inspect the engineering works for quality and general compliance with all requirements

(iv) Assist the Contractor in evaluating the subcontractor's programme and progress, in respect of design work, procurement, installation and commissioning

(v)   Witness tests

(vi) Assist in ensuring that the subcontractor complies with supplementary Contract requirements, including reports, sanction procedures, as-built drawings, maintenance manuals etc.

(vii) Provide advice with respect to payments and variations etc.

(viii) Provide advice in cases of conflict between the Contractor and the Employer, or between the Contractor and the subcontractor as required

(ix) Assist in resolving deficiencies or defects which may arise during of after completion of the Works.

Consultant-designed Systems

When a contractor requires the building services engineer to carry out the design of some or all of the engineering work, he may use the following check list to define the services:

Paragraphs in italics may not be required if the work was undertaken satisfactorily during the pre-contract stage.

(i)   Prepare, or amend, any drawings or details needed for incorporation into the Contract Documents (ii)　*Examine with the Contractor and other members of the design team, all available information in order to establish the building services requirements of the Works*

(iii)　Assist in the preparation of the design and information release programme

(iv)　*Visit the site and make enquiries to determine any factors which could affect the engineering design*

(v)　*Advise on the need for any surveys*

(vi)　*Consult the Local or any other relevant Authority or Statutory Under-taker on matters which may affect the design of the engineering work*

(vii)　*Advise on the need for any other specialist consultant's advice in connection with the engineering works*

(viii)　*Liaise with the Contractor to determine the scope of the engineering works to be sublet to specialist subcontractors, and provide all information which will be required to enable tenders to be obtained from such subcontractors. Assist in the evaluation of tenders received*

(ix)　Provide the Architect with information which he needs to coordinate the design of the works

(x)　*Advise on the need for any tests that may be required,* provide the necessary instructions, and as necessary witness and report upon such tests

(xi)　Provide information necessary and assist as necessary in obtaining approvals and sanctions required by the terms of the Contract

(xii)　Provide the Architect or Structural Engineer with any information required in relation to builder's work in connection with specialist sub-contractors' works

(xiii)　Examine specialist subcontractors' proposals and working drawings for consistency with the intent of the engineering design and specification

(xiv)　Prepare working drawings, details, specifications and schedules which the Contractor or subcontractor will require to install the building services

(xv)　Inspect the engineering works either by visiting, or by employing a resident engineer

(xvi)　Comply with supplementary Contract requirements in respect of the building services, including reports, sanction procedures, as-built drawings, maintenance manuals, etc.

(xvii)　Attend meetings as required by the Contractor

(xviii)　Assist in practical and final completion inspections of specialist sub-contractors' engineering works

(xix)　Assist the Contractor by providing technical advice in the event of any dispute arising, either with the Employer, or with any building services subcontractor.

CONTRACTORS' DESIGN MANAGEMENT

As Design–Build contracting has become more common, the role of the contractor's design manager has become increasingly recognised as a role in its own right, and not a role for the consultants nor a secondary role for contractors' contract management staff.

As most contractors have their own systems and procedures, and as all projects are unique, it is difficult to define the role of the design manager, in anything other than a generalised way. This can be done by referring to a check list of duties that may be assigned to the design manager, as follows:

(i) Establish the need for consultants, prepare and complete the agreements with them, including the definition of the scope of the works, the services required and the terms which are to apply. Ensure that the consultant's terms are compatible with the requirements of the Contract

(ii) With the aid of the consultants prepare an information production programme, taking into account the needs for coordination between the various disciplines, sanction procedures, procurement and construction

(iii) Obtain, from the consultants, information necessary for seeking quotations from specialist subcontractors

(iv) Ensure that orders or instructions are given to specialist subcontractors in adequate time to allow the consultants to incorporate the subcontractors' details etc. into the design of the works as a whole

(v) Liaise between the design and the construction teams to ensure that each understands the needs of the other

(vi) Liaise with the Employer to ensure that his requirements are met, be they included in the Contract, or be they changes

(vii) Liaise with the Contractor's in-house cost control staff to ensure that the designs, as they develop, do not entail on the Contractor unnecessary additional costs

(viii) Liaise with the construction staff to ensure that the designs, as they are developed, take due account of buildability and availability of materials

(ix) Monitor the progress of the design team

(x) Monitor applications for Statutory Approvals

(xi) Continuously monitor the performance of the design team against all the requirements of their appointment, including site inspections

(xii) At completion and commissioning ensure that all the Contract requirements are met in terms of certification, as-built drawings, manuals, etc.

(xiii) Work with the Contractor's cost control staff in settling the Final Account, and subcontractors' accounts

CONTRACTOR'S CONSULTANTS' FEES

Invariably, contractors prefer to agree, with their consultants, lump-sum fixed fees inclusive of all disbursements, subject to variation only if there are substantial variations to the work which involve the consultant in a significant amount of extra work. Most consultants cooperate with the contractor on this matter.

Pre-Contract Fees

Architects and Engineers

Generally architects and engineers agree to provide their pre-contract services on a 'no-job-no-fee' basis, although contractors occasionally agree to make a nominal contribution towards consultants' costs, in particular in respect of presentation materials.

Such lump-sums that are paid, are usually negotiated between the parties and bear no relationship whatever to fees as recommended by the various professional institutes.

Professional Quantity Surveyor

If a professional quantity surveyor agrees to provide his services on a no-job-no-fee basis, or at a reduced rate, it is normally in consideration of an enhanced fee, should the tender be successful, to reflect the speculative nature of the assignment. It is impossible to make generalisations on the split and level of enhancement of such fees.

Where the fee is not complicated by such a split, experience shows that typical charges by recognised professional quantity surveying practices for the preparation of bills for Design–Build tenders are either based on an hourly rate or lump-sum basis, as follows:

(i) Hourly Rates
Hourly rates usually fall between
£1.50 and £1.70 per hour per £1,000 of annual *salary*

(ii) Lump-Sum Charge
Lump-sum charges for bill production vary considerably, and generally bear little relationship to charges suggested by the RICS for professional quantity surveying services.

Charges quoted by recognised professional quantity surveying practices for the preparation of detailed bills of quantities, largely to a standard method measurement, are usually
Not less than £4 000 for a £400 000 project
Not more than £50 000 for a £10 000 000 project
This implies a sliding scale of 1 per cent to 0.5 per cent, but in

practice, charges can be as little as 0.25 per cent for straightforward or repetitious projects, or for projects where the contractor asks for a less detailed bill of quantities.

Freelance Quantity Surveyors

Freelance quantity surveyors are often prepared to produce bills of quantities for charges of £100s not £1,000s. For simple projects this may be a good and economic way to have bills prepared, but contractors must be cautious, and ensure that they make thorough cross-checks of such bills before tendering.

Fees for Post-Contract Services – Contractors' Consultants

Here we look at the fees that contractors might expect to be charged for post-contract services, for the different varieties of Design–Build, and for buildings of different complexities.

The fees suggested are in very broad bands because of the complete impracticality of preparing charts or tables similar to those issued by the various professional institutes.

Contractors' Architects' Fees

Fees are normally agreed between contractors and architects without any specific reference to the RIBA scales, nor even to the precise contract sum.

Regular Design–Build contractors and architects have their own ideas for an acceptable range within which to pitch the fees. Generally they will not fall outside the following bands:

(i) Contractor-led-design (full service as per this chapter)
   (a) Simple repetitious building   2%–3%
   (b) Complex building   3%–5%

(ii) Employer-led-design (full service as per this chapter)
   (a) Simple repetitious building   1%–2%
   (b) Complex building   2%–3½%

These percentages must be read in conjunction with the following points:

(i)   The percentages are to be applied to the contract value, after ducting the cost of all fees.
(ii) The charges would be inclusive of all pre-contract costs.
(iii) The charges would be inclusive of all disbursements.
(iv) The buildings envisaged would range in value between say, £400 000 and £10 000 000. Outside these values, different percentages would apply.

(v) The percentages relate to new work. For alteration work, a further 1 per cent to 2 per cent may be added depending upon the nature of the work.

Contractors' Structural Engineers' Fees

The post-contract services provided by a structural engineer are similar for both Design–Build and Traditional Contracting, and thus reference may be made to the ACE fee scales.

The percentage calculated from these scales is applied to the cost of the engineering works, as defined in the ACE Conditions of Engagement. This can pose some difficulties, not least, because the break down of the project cost is not usually available when the fees are agreed.

In practice, the engineer makes his own assessment of the value of the engineering works, calculates the fees himself from the scales and, more often than not, offers the contractor a discounted rate. The fees thus quoted usually lie between 60 per cent and 80 per cent of the scale fees, depending on the fullness of the service required and the nature of the project. The experienced Design–Build contractor will usually know whether the quoted fees are competitive, and if he thinks that they are too high, he will normally negotiate with the engineer.

Contractor's Building Services Engineer's Fees

The services required of building services engineers vary so much from project to project, and from contractor to contractor, that it is difficult to make any firm suggestions for general fee levels.

However, in the absence of any other source, reference may be made to the ACE Conditions of Engagement 4A (iii) for some guidance on fee levels, as the services described therein are largely similar to those which many contractors would require on projects where the services are complex or predominant.

From these conditions it can be deduced that the fees for providing the service as described earlier in this chapter as 'subcontractor designed systems' should be: between 3 per cent and 4 per cent of the cost of the engineering works.

If the consultant is to design the systems, then reference should be made to the ACE Conditions of Engagement 4A (i) for applicable percentage fees, and as with the structural engineer, contractors can expect to be offered a discounted rate.

17 Valuation of Changes

'Change is not made without inconvenience,
even from worse to better'

Richard Hooker [*Quoted by Johnson*]

KEY TOPICS

- Source of changes and variations
- Changes affecting current work
- Contractor's firm quotation for a change:

  consultant's involvement
  estimate of cost:
  measured work
  specialist subcontractor's work
  preliminaries
  fees
  profit

- Documents describing the proposed change
- Programme
- Proceeding with changes before agreement on cost or time
- What if a proposed change does not go ahead?
- Contractor's actual costs

INTRODUCTION

This topic, the cost of variations, is an important one to understand, as the alleged lack of control over these costs by the employer is often cited by antagonists of Design–Build as one of the greatest drawbacks of the system.

We have seen in Chapter 4, 'Forms of Contract', the eventualities which can lead to a variation in the contract sum, and how they may be within or beyond an employer's control.

In this chapter we concentrate on variations brought about by changes in the Employer's Requirements, to use JCT 1981 terminology, or Client's Representative's Instructions, to use BPF terminology. In either case we shall use the word 'change' to mean the change in the requirements, and 'variation' to mean the adjustment to the contract sum payable to the contractor.

Before considering the subject in detail, we shall first consider the definition of the various 'costs' connected with a change, namely:

(i) the contractor's estimated cost,
which is the contractor's estimate of the adjustment to the cost to himself of completing the works as a result of the change.
(ii) the contractor's quotation,
this is the amount which the contractor quotes for the change, based upon the provisions of the contract, i.e., based on rates contained in the schedule of rates – or similar provisions depending on the particular contract terms, together with appropriate amounts for design and profit.
(iii) the cost to the employer of the change,
the amount that the employer and contractor agree upon for the variation.
(iv) the cost to the contractor of the change,
the amount that it actually costs the contractor to carry out the change.

Sources of a Change

The instigator of change a may be an employer when his needs change, or when he wishes to make an improvement to the works. Alternatively, a proposal may come from the contractor or his consultants for a change which may be of benefit to the employer in one way or another.

Normally, a change will add cost to the contract, and possibly add time to the programme. However, it must not be forgotten that some changes may be conceived with the aim of reducing the cost, or time.

Variations may also arise from changes that are necessitated by changes in statutory requirements, but it will depend on the contract terms whether the contractor will be reimbursed his additional costs for such a change. For the purpose of this chapter we are only considering changes which do rank for reimbursement.

What if the Proposed Change Affects Current Work?

When a proposed change affects current work, the contractor should make this fact clear to the employer, and between them, they will have to agree the best approach. Unless the employer agrees otherwise, a contractor should never hold back progress in anticipation of a change without instructions from the employer. None of the standard forms of contract provides for an extension of time for delays incurred by a contractor waiting for an instruction to make a proposed change if, in the event, no such instruction is ever forthcoming.

Generally, such situations are resolved by good communications between contractor and employer, and it is up to the contractor to ensure that he keeps the employer fully informed of the consequences of whatever action they jointly agree. If, at the time, the contractor is unable to estimate and foresee the full consequences of a proposed change which will affect current works, he should also make this clear to the employer.

The employer may instruct that the change should be made, on the understanding that the cost, and effect on the completion date will be agreed later. A firm date may be agreed for the contractor to provide this information to the employer.

It may seem a loose way to deal with the costing of a change, but it is not the contractor, nor the Design–Build system, but the circumstances which cause such a problem; no less than they would in any other form of contracting.

Firm Price before Proceeding, or Later Negotiation

In the event of a change being envisaged, the employer will either demand a firm proposal in terms of design, cost and price before he will sanction or give instructions for a change; or time may not permit this, in which case he will give his instructions for the change in whatever way suits him best, or best protects his interests. This may be by agreeing with the contractor 'a guaranteed maximum' in terms of price and time, or if even this is not possible, the parties may agree that the contractor will proceed with the change, and that it will be valued subsequently by negotiation.

CONTRACTOR'S FIRM PROPOSALS FOR A CHANGE

The contractor will have to prepare firm proposals whether they are required before instructions are received from the employer to proceed with the change, or afterwards.

Action upon Receipt of a Request for a Change

When a contractor is notified of a proposed change, he should agree with the employer the manner in which the he should respond, both in format and time.

Clearly, if a proposed change is speculative on an employer's part, the contractor should be asked only for outline proposals and no more than a budget price. This will enable the employer to test the feasibility of the proposal before committing the contractor or himself to an expensive exercise.

If, on the other hand, the proposed change is 'serious', then the contractor will normally be asked for detailed proposals with a firm price and break-down, and a firm statement on the effect on the programme.

Who, in the Contractor's Organisation, Estimates the Firm Price?

When the contractor receives a request for proposals and a firm price for a proposed change, he should deal with it in the same way as he dealt with the preparation of the original tender. Once a firm price and programme is given by the contractor and accepted by the employer, there will be no going back for more later. In practice an estimator prepares the original tender for a contractor, and often upon receipt of the order, he hands over the documentation and responsibility for financial control to the contractor's post-contract quantity surveyor.

In Traditional Contracting, a contractors' quantity surveyors normally prepare and negotiate variations with employers' professional quantity surveyors, and they often do likewise on a Design–Build project. It is here that contractors must be careful, and ensure that only quantity surveyors, who have the requisite experience and ability, prepare estimates for changes which involve 'interpretation' of incomplete designs and quantities.

It is important therefore for contractors to ensure that only a suitably competent person prices the change.

Involvement of the Consultants

It is tempting for contractors to think that they can deal with changes without involving their design team. This is dangerous. If an architect is appointed on the basis that he is the design team leader (and good practice dictates that normally he should be), then he is responsible for design coordination. He, the architect should therefore be advised of the proposed change, and he or the contractor should ensure that all the other designers, including specialist subcontractors, are notified. Each must have the opportunity to consider whether the change could or would affect their own work.

The design development for a change must be coordinated in the same way as tenders are.

Contractors normally rely upon their consultants to ensure that the proposals take into account all statutory requirements, and indeed to take into account all factors which could affect the design.

The Contractor's Estimated Cost

Contractors' estimates are in five parts:

   (i) measured direct work, or non-designer subcontract works,
   (ii) works by a specialist subcontractors, who are already in place,
   (iii) works by a 'new' specialist subcontractor,
   (iv) preliminaries,
   (v) fees.

Measured Work – Contractor's Estimate

When the designs for a change are sufficiently advanced, the contractor may prepare a form of bill of quantities, and price it himself and/or obtain quotations from subcontractors and suppliers. His objective is to estimate the cost to himself of the work.

Specialist Subcontractors, Already in Place, Contractor's Estimate

When a specialist subcontractor is already in place for the proposed work, he will be involved in the design process and from that determine the effect upon his own work. He can then price his work and provide the contractor with whatever back-up details are necessary to justify the price to the contractor in the first place, and the employer in the second.

Contractors and these subcontractors, should ensure that any changes in attendances, or builder's work in connection with the subcontractor's works, are taken into account.

If necessary, subcontractors will describe or illustrate their proposals by way of drawings and specification.

'New' Specialist Subcontractors, Contractor's Estimate

Where there is no subcontractor in place for proposed additional work, contractors may have a realistic opportunity to obtain competitive quotations, if time and circumstances permit. However, the process of preparing a performance specification, and enquiry documents to obtain competitive proposals is often far too time consuming to be contemplated for establishing an estimate for a change. The alternative is to go direct to a single known subcontractor, and to discuss design proposals with him so that he can prepare a quotation relatively quickly.

This may not appear to produce the lowest possible price, but the discussions with the specialist subcontractor can often produce a more economic or 'better' solution than would have otherwise been devised.

Preliminaries, Contractor's Estimate

Contractor must consider the effect of the change on their programme, and consequently their preliminaries, and there can be difficulties here. A contractor may quote his price and programme based on a 'latest date' for the instruction to proceed, and if the employer does not meet that date, yet gives his instruction later in the form of 'I accept your quotation and your revised completion date', it is an invalid instruction.

However, what does the contractor do? If he rejects the instruction further delay is caused and the cost rises still further. This situation can go round in circles, each party blaming the other for the unsatisfactory state of affairs. The only way that the contractor can protect his interests is by being open, firm and reasonable, and making sure that the level and quality of communication he has with the employer is to such a standard that such arguments never arise.

Design Fees, Contractor's Estimate

The terms of the agreements which the contractor has with his consultants will determine the fees in respect of changes. Subcontractors' design fees will normally be included within their quoted prices, so the contractor need make no further provision for them.

Fees to Statutory Authorities, Contractor's Estimate

It is unusual for a change to involve any additional costs in respect of Planning or Building Regulation fees. This would be a matter which the architect could advise the contractor upon but, in any case of doubt, the Local Authority should be consulted.

Profit, Contractor's Estimate

Contractors will wish to include profit at a level which at least equals the level included within the original tender, and normally at an enhanced level. However, for the purposes of the estimate they would normally apply a reasonable percentage to estimated costs, and thereby arrive at a sum which they would regard as the minimum acceptable price for the change.

Preparing the Quotation for the Change for Presentation to the Employer

Once a contractor has estimated what it will cost him to carry out the change, and the minimum profit that he would find acceptable. He has to calculate what he should quote the employer for the change. It is in this field that the contractor's quantity surveyor is, or should be, expert. He will conform to the terms of the particular contract to establish the best possible price for the work.

This is the aspect of Design–Build contracting which can cause the most serious disagreements between employers and contractors, and so it is up to contractors to make sure that their quotations are presented in the most acceptable way, and negotiated in an open and straightforward manner, backed-up by whatever detail and information an employer could reasonably expect.

Contractors are normally aware that any future work with the employer will depend, among other things, upon the contractor's approach to valuing changes. Furthermore, Design–Build contractors normally have to rely upon references and recommendations to obtain work with new employers and so, again, impressions are important.

The details of the quotation will normally follow the same format as the contractor's estimate, but it will be put differently; as we shall see.

Measured Work in the Quotation

The contractor may use the bill of quantities which he has already prepared as the basis of this part of his quotation, but the description of the items and the rates which would apply are not necessarily those which he used for his estimate, but those which are contained in one form or another in the contract documents, e.g., in the Contract Sum Analysis for a JCT 81 contract, or the Schedule of Activities for a BPF contract.

Specialist Subcontractor's Work in the Quotation

Often with specialist subcontractors' work, there will not be any item in the contract documents which can be used as the basis for measuring and valuing changes. If there is then that can be used in the same way as the measured work is presented. If there is no such similar item, as would be the case, if the change was, for example:

> the introduction of air conditioning to an additional room, or
> an increase in the lighting intensity in a particular area,

then the valuation of this part of the change would be by presentation of a detailed build-up of the price, incorporating where possible, suppliers' prices for materials, plant or equipment.

When the sub-contract work is to be by a 'new' subcontractor, and again there are no comparable rates within the contract documents, for example, if the change was say,

> introduction of specialist joinery in up-grading the finish in a reception area, or
> introduction of a new services element, such as an audio-visual system,

then the contractor would include, in his back-up, a copy of quotations that he had received for the work.

Preliminaries Included in the Quotation

If there is anything within the contract documents which lays down the manner in which preliminaries are to be valued in the event of a change, then that is the way in which it would be done. Otherwise, the contractor will give a detailed build-up of his additional preliminary items; either extra items or extended durations for established items.

Contractor's Quotation for Design Fees on Additional Work

A frequent omission from contract documentation is the manner in which design fees are to be calculated for changes. In the absence of specific terms to the contrary, the traditional format is usually adopted. That is, a percentage is added to the value of the change. The actual percentage rate would normally be the same as that included within the original contract sum. There will be many instances where this is patently inappropriate. For example, on the one hand, when the change is merely the upgrading of the specification of a particular material, in which case little or no additional design work would be involved, or on the other hand, when the cost of the design work could be far more than the a percentage of the cost of the change.

The contractor and employer will have to agree how to resolve these inequalities. They may decide to resort to 'reasonable' negotiations in each case, or stick to a percentage, on 'swings and roundabouts' basis.

It should be noted that it is for the contractor to manage the possible difference between his consultants' terms and the terms which he has agreed with the employer.

Profit on Additional Work, Contractor's Quotation

The Contract Sum Analysis, or its equivalent, will contain the declared profit and contribution to overheads which is included in the contract sum. The percentage which this represents is the percentage which would normally be applied to the additional net cost of changes. Few employers or contractors would disagree with this principle. If either did disagree in any particular case he would have to produce convincing arguments to support his view.

Design Fees and Omissions or Substitutions of Work

Generally, for the purpose of establishing the value of design fees, especially in respect of omissions, it is most simple to regard design as 'work' in the same sense as site work. Thus, if at the time of the omission of site work, the design work in respect of it was already complete, then there should be no reduction in the design fees. Furthermore, if an omission is cancelled out by an addition, then not only should there be no reduction in relation to the omission, if the design work was completed, but there should also be an addition to the fees payable on account of the additional work calculated in the normal way.

Profit and Omissions

Whilst the standard forms of contract specify the way in which the omitted works themselves are to be valued, they are not specific in respect of loss of profit. It may seem reasonable and equitable that the profit, which is often regarded as 'percentage' on-cost, should also be reduced in proportion.

The argument against this approach is that the profit is a single lump sum quoted in competition, or negotiated, and as such is inviolable. This is as opposed to additional work, where the contractor did not contract in the first place to undertake the additional work and therefore if he is to take on the extra work the same rules for profit should apply as they did in the original work; that is, a similar percentage should be added for profit.

The argument is largely academic, because in practice, it is exceedingly rare to have a final contract sum less than the original contract sum, because invariably any omissions are outweighed by additions. This being the case, the profit will be calculated on the net addition, and not on the additions, without a corresponding reduction on the omissions.

Contingencies in the Contractor's Quotation

Contractors will always want to include a contingency in the pricing of the change commensurate with the risks involved. (The risks are the same in essence as those which he takes on in the project as a whole, see Chapter 11.)

However, it is not normal to add a specific contingency in the build-up, other than for the employer's use or risk. The reason for this is that it is difficult to obtain the employer's agreement to include money to be paid to the contractor 'for nothing', which is what the contractor's contingency would be if no unforeseen extra cost arose; or if an extra cost arose which the contractor had not foreseen, then the employer's normal view would be that it would be the contractor's mistake, and why should he the employer pay for contractor's mistakes.

So, any contingency which the contractor wishes to include must be contained within the rates established in the first place, or within his profit margin.

Documents Describing the Proposed Change

Contractor should prepare sufficient documents by way of drawings, details and specifications to illustrate his proposals detailed to an extent that is appropriate to the change proposed. The contractor must be careful to include details of any secondary changes consequent upon the principle change. They should describe the works fully so that no conflicts arise later.

'Current' drawings should not be altered, until instructions are received from the employer, in case the change is not accepted.

Before he present his proposals to an employer, the contractor should go through a routine of eliminating discrepancies, omissions and errors using the same principles that apply to Design–Build tendering in general.

Programme

Contractors must consider the effect on the programme. If there is plenty of time to consider a change, then it should be relatively simple to establish whether an extension of time will be warranted, or necessary. In practice, most changes arise late, with little or no time available before implementing them. So it is often quite difficult to establish the effect on the programme, and to compound the problem it is often affected by the date on which the employer gives the go-ahead.

It may seem simple for the contractor to quote programme times, or effect, based on a 'latest date' for the instruction from the employer to proceed, but in practice no matter how firm the contractor endeavours to be, the employer will very often try to get the contractor to agree not to extend the programme, even if he fails to give instructions in good time. An argument sometimes put forward by employers is that: 'if the contractor had provided his price and proposals more quickly, then there would not have been a question of delay'.

So, it is not only important for contractors to carefully work out the consequences of proposed changes on their programmes, but also to programme the production of quotations for changes, and make due allowances for the employer to consider the proposals, prior to the instruction to proceed. If there is insufficient time for this process then contractors should inform the employer at the outset and let him, the employer, make the decisions.

PROCEEDING WITH A CHANGE BEFORE AGREEMENT ON COST AND TIME

Ideally, the production of firm proposals described earlier in this chapter should be presented and approved prior to the implementation of the change, but frequently there is insufficient time to do this. In this case the employer will give the go-ahead in a form appropriate to the circumstances at the time.

The contractor will satisfy himself that his position is protected before he accepts the instructions to proceed with the design, procurement and the site work in accordance with the required change.

At an appropriate stage later, the contractor will produce his firm proposals, as described earlier, and the variation to the contract sum, and extension of time, if applicable, will be negotiated and agreed.

If the contractor and employer fail to agree on the value of the change, arbitration is a remedy, but this rarely profits either party, and so strenuous efforts should be made to avoid this last resort.

The BPF/ACA Form, and the JCT 81, BPF Supplementary Provisions, however, provide for the employment of an Adjudicator. This is quicker and less drastic than going to arbitration, but even so it would mark a serious deterioration in relationships between the parties, and again should be avoided if at all possible.

Apart from presenting his quotation in a clear and detailed way, the contractor's other way to reduce the likelihood of serious disagreement, is to make certain that all instructions are recorded, whether they are given formally or otherwise; that he keeps a record of all discussions between the employer and any members of the contractor's staff or design team; and that a record is kept of all hours and materials expended in relation to the change, to support any subsequent submission he may have to make to the employer.

WHAT IF THE CHANGE DOES NOT GO AHEAD?

Contractor's Costs in Making Proposals

When a proposed change is likely to involve a contractor in considerable work before he can make firm proposals, he should be aware of the costs which he will incur in making the proposal, and unless the contractor specifically agrees with the employer that the employer will pay for such costs, then the contractor is unlikely to obtain reimbursement of his costs if the change does not go ahead.

The particular contract conditions may include, or exclude provision for such payments, but the common forms are silent on the subject.

This does not prevent a Design–Build contractor from treating the design work resulting from a request from the employer for a change as a 'change' in itself. As we have already seen, it makes the position more clear if design work is considered as 'work' in the same sense as site work.

Thus, a contractor could quote fees, for preparing the proposals, which the employer may accept or reject. If the employer accepts the fee quotation, all well and good.

On the other hand, the employer may not accept the fee quotation either because he decides not to proceed with the change, as may be the case if the proposed change was highly speculative; or he may still instruct the contractor to proceed with the necessary design work, and decide to agree a 'fair valuation' for the design work at a later date, in the same way that he would with a disputed quotation for site work.

Basis for Agreement of Costs of Preparing Proposals

In practice there are fewer disputes than may be supposed over abortive costs for preparing proposals for changes that do not go ahead. Usually the costs are resolved in one of the following ways:

(i) If the work involved is relatively small, or if the change is originally proposed by the contractor, then the contractor may waive any abortive charges

(ii) If the change is brought about by a change in statutory requirements, then it will go ahead, and the question of abortive costs for preparing the proposals will not arise

(iii) For additional work, the employer may accept the contractor's design costs either on an 'hourly rate' basis, or on the basis of a percentage of the cost of the work involved, as if the work did proceed, provided this does represent a fair and reasonable amount

(iv) For the omission or the substitution of work, it is either difficult or even impossible to make an assessment of a fair charge for abortive design work on a percentage basis; a fair and reasonable assessment could be made on an hourly rate basis

(v) A contractor does not normally make a charge for his internal non-design costs.

Naturally contractors should ensure that their agreements with their consultants are on a back-to-back basis. That is, a contractor should protect himself from incurring abortive design cost that he is unable to recover from the employer.

THE CONTRACTOR'S ACTUAL COSTS

Once an employer and contractor have agreed the variation, the contractor will have no grounds for further payment should he have underestimated the work, or not appreciated consequential costs arising as a result of the change.

Hence, it is only the contractor who is interested in the actual cost of the change; his profit or loss on the variation is the value of the variation less the actual cost to him.

IV   Related Topics

18 Value Engineering and Design–Build

'This was the most unkindest
cut of all'

Shakespeare [*Julius Caesar*]

ORIGINS OF VALUE ENGINEERING

Value engineering is a technique which originated in the engineering industry in the USA after the Second World War to improve efficiency of manufacturing processes.

It has spread into the construction industry and slowly out of the USA to other countries, including the UK. In the UK it has hitherto been practised mainly by consultants who have trained in America, or by firms with American connections.

WHAT IS VALUE ENGINEERING?

Value engineering is a technique that questions and examines, in a structured way, every aspect of a project in an endeavour to seek improvements, which may be in capital cost, running and maintenance costs, efficiency of spatial planning, function, appearance, return on investment; in short, anything which will give the employer better value for money.

Value engineering exercises may be undertaken in one or several stages during the inception and design development of a project. The balance must be made between approaching the studies too early, in which case there may not be enough basic data evolved to examine productively, or too late when any change will be too costly to implement.

Value engineers break down the task into stages, for example:

(i) establishing the brief and objectives
(ii) studying the current proposals
(iii) systematic questioning of every aspect of the project as proposed, and making counter-proposals
(iv) evaluating the options which have arisen
(v) translating the options into consolidated proposals and putting them forward for the employer's approval

The depth of the exercise will vary depending upon the stage that the design has reached. Different terms are applied to stages, as follows:

Briefing stage	– Challenge brief
Design development stage	– Full Value (VE) exercise
Tender evaluation stage	– Value audit
Construction stage	– Value audit

WHICH PROJECTS BENEFIT FROM VALUE ENGINEERING?

In theory, every project could benefit from the application of value engineering, but it is probably only those which would be generally regarded as large, complex or expensive, or those which are in of a series of repetitious projects, which would justify, or benefit from, a 'professional' value engineering study. On straightforward work the law of diminishing returns may apply.

Others in the construction industry will often be called upon to undertake a study, which some may describe as value engineering, when the anticipated cost of a proposed project exceeds the employer's budget. In such cases an architect, quantity surveyor or a contractor will often be asked to look at the job to seek savings. Such exercises are more appropriately termed 'cost-cutting' exercises, and not value engineering.

EFFECT OF THE CHOSEN FORM OF CONTRACT

Traditional and Management Contracting

Value engineering can be applied conveniently to the design in both Traditional and Management Contracting. During the pre-tender stage the study, or a series of studies can be implemented, often to beneficial effect. With this form of building procurement there would otherwise be no party, other than the employer himself, who has a direct and absolute interest in questioning design decisions.

The architect and the structural and building services engineers all make their own design proposals and, in practice, not one seriously challenges the work of the others. The professional quantity surveyor has a direct interest in seeing that any budget he has put forward is not exceeded without reason, but if budgets are not exceeded then, generally, he will not make comment in a true value engineering sense.

DESIGN–BUILD CONTRACTING AND VALUE ENGINEERING

We have seen in Design–Build that the design of a building is generally divided between the employer's consultants and the Design–Build contractor or his consultants.

Employer's Design

During the stage in which the design is being developed by the employer's consultants, in a Develop and Construct or a Design–Build (single-stage tender) project, a value engineer could be employed to question the brief in the first place, then to implement a full value engineering study of the employer's consultants' proposals before the enquiry documents are prepared. See Table 18.1.

In a contractor-led-design project the value engineer's work, in relation to the employer's design, would be restricted to questioning the brief only. See Table 18.2.

Contractor's Design

The situation is fundamentally different when it comes to a contractor's design in a Design–Build contract. The essence of Design–Build is that contractors' proposals are judged not on price of the construction work alone, but also on the design and the value for money that flows from it.

During a contractor's design process his one aim is to produce 'the best scheme' for the employer so that he will be awarded the contract. The better the scheme, and the better the value for money, the more likely is the contractor to be successful.

The successful Design–Build contractors, therefore, apply value engineering techniques, in one form or another. As yet, few contractors employ specialist value engineers, and still fewer have a 'value engineering department'. Inevitably, the number will increase, and those that lag behind will find that they become increasingly less successful in their Design–Build proposals.

However, the contractor's opportunity to improve the design increases with the proportion of the design that is allotted to him. Thus he has more scope in contractor-led-design projects than in employer-led-design projects. See Tables 18.3 and 18.4.

Table 18.1
Employer-led-design – Value engineering of employer's design

Concept	Brief	Design Devmnt.	Tender	Construction
–	Challenge brief by employer	Possible full VE by employer	–	–

Table 18.2
Employer-led-design – Value engineering of contractor's design

Concept	Brief	Design Devmnt.	Tender	Construction
–	–	Doubtful by contractor	Value audit by employer	Value audit by contractor

Table 18.3
Contractor-led-design – Value engineering of employer's design

Concept	Brief	Design Devmnt.	Tender	Construction
–	Doubtful by employer	–	–	–

Table 18.4
Contractor-led-design – Value engineering of contractor's design

Concept	Brief	Design Devmnt.	Tender	Construction
–	–	Possible full VE by employer or contractor	Value audit by employer or contractor	Value audit by contractor

19 Collateral Warranties

'..to set his seal, to give the world
assurance of a man'

Shakespeare [*Hamlet*]

INTRODUCTION

This book is a practical guide to the use of Design–Build and, as such, it does
not seek to provide solutions to complex legal issues. However, the author
felt that the publication would not be complete without some mention of the
vexed question of collateral warranties.

In this short chapter we see why the demand for collateral warranties has
grown so dramatically in the late 1980s, and we look for hidden dangers to
contractors and consultants in giving warranties.

It must be added that we are not dealing here with criminal issues.

THE LAW AND DEFECTS

In Contract

In simple terms, if there is a contract between two parties, and one fails to
fulfil his obligations under the contract, then the other party may seek
recompense from the 'guilty' party and if necessary sue him in the courts for
damages. This is what an employer is able to do if a Design–Build contractor
produces a building with defects or deficiencies in contravention of the
terms of the contract.

Certainly an employer suffering in this way, would be able to recover all
direct costs, loss and damages, and quite possibly indirect and 'economic'
losses.

So where there is a contract between the parties in dispute it should be
relatively simple to establish both liability and the level of compensation to
be paid.

In Tort

Where there is no contract between disputing parties, the issue becomes
tortious. That is, it is a non-contractual, civil wrong giving rise to action in
the courts.

Court Findings

Recent court decisions on the question of liability, in the first place, and the extent to which damages are recoverable in the second place, have swung like a pendulum, leaning first towards the plaintiff, then towards the defendant. It is not appropriate, in this book, to go into any detail on this point, partly because it is not intended to be a guide to the law, and partly because the situation is ever-changing.

What can be said with certainty is that tortious issues are clouded with uncertainty, and a plaintiff is less likely to succeed with a claim for damages in tort than he would in contract.

Situations where Tortious Actions could Arise

Given a defective building, there are a number of circumstances which could give rise to a situation where a plaintiff has no contract with the defendant, and thus his court action is tortious, and not contractual. These include, for example:

    (i) When the employer sells the building (and the consequent owner sells on again, ad infinitum).
    (ii) If a fund sells their interest to a second fund, and so on.
    (iii) If an employer goes into liquidation and the building reverts to a guarantor, or fund.
    (iv) If a lessee or tenant suffers 'damage'.
    (v) If the contractor goes into liquidation, and the plaintiff, be he the employer, a tenant, or fund, seeks recompense from the designers (assuming that they were not the contractor's direct employees).

Collateral Warranties and Design–Build

Collateral means alongside or parallel, and a warranty is an undertaking on a vendor's part that his service will fulfil specified conditions.

Hence, in respect of building design, collateral warranties are given by contractors, and designers if they are separate bodies from the contractor, to parties nominated by the employer, for example a fund, tenant or prospective purchaser. Furthermore the employer may seek a warranty from the designers. Figure 19.1 shows the complex network of warranties which could sought by an employer where a fund and tenants are involved in a project.

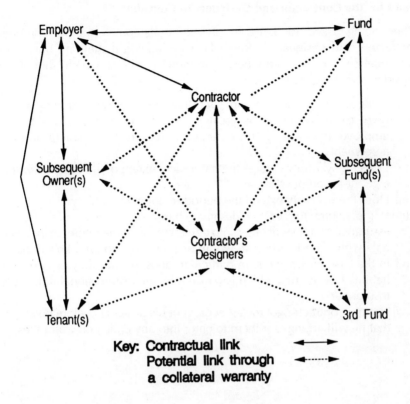

Figure 19.1   Contracting parties and potential parties to collateral warranties

Nature of the Warranty

There are few standard warranties in common use, and even if there were, experience shows that employers and funds would be most likely to produce their own to suit their own particular legal advisor's views.

Usually warranties are designed to give funds, purchasers and tenants the same benefits that an employer enjoys under the terms of the contract with the contractor; and the same benefits that a contractor enjoys in his agreement with his designers.

Contractors and designers do not normally enter into warranties with subsequent owners tenants and funds at the time of the main contract, because they are not known. Normally, at the time of the contract the contractor and designers are asked to agree to a contract which obliges them to enter into the requisite warranties as and when tenants are known, and when and if the building is sold on, or the fund sells on its interest; both these latter eventualities, ad infinitum.

Issues for the Contractor and Designers to Consider

When being asked to agree to provide collateral warranties, prudent contractors and designers should take legal advice if they are unsure of consequences of the warranties. Common questions which should be considered include the following:

(i) Is the nature of the warranty such that the duty the warrantor would owe to the warrantee is greater than the contractor owes to the employer in contract, or the designer owes to the contractor in their agreement?

(ii) Does the warranty change the nature and extent of the design liability, e.g., 'fit-for-purpose'?

(iii) Does the warranty extend the period of potential liability?

(iv) Is the contractor, or the designer, happy to give a carte blanche assurance that he will enter into agreement at some future unspecified date with a third party with whom he may have no prior knowledge?

(v) In the case of designers, in particular, does the warranty extend their liability beyond that which is covered by their professional indemnity insurance?

(vi) Has the contractor got the agreement of his designers before he agrees that he will arrange for them to enter into any collateral warranties?

Index